T0317612

PROCESS STEAM SYSTEMS

PROCESS STEAM SYSTEMS

A PRACTICAL GUIDE FOR OPERATORS, MAINTAINERS, DESIGNERS, AND EDUCATORS

Second Edition

CAREY MERRITT

Registered Offices
John Wiley & Sons, Inc., 111 River Street, Hoboken, NJ 07030, USA
John Wiley & Sons Ltd, The Atrium, Southern Gate, Chichester, West Sussex, PO19 8SQ, UK

Editorial Office
111 River Street, Hoboken, NJ 07030, USA

For details of our global editorial offices, customer services, and more information about Wiley products visit us at www.wiley.com.

Wiley also publishes its books in a variety of electronic formats and by print-on-demand. Some content that appears in standard print versions of this book may not be available in other formats.

A catalogue record for this book is available from the Library of Congress

Hardback ISBN: 9781119838937; ePub ISBN: 9781119838951; ePDF ISBN: 9781119838944; oBook ISBN: 9781119838968

Cover image: Courtesy of Attis Co.
Cover design by Wiley

Set in 10/12pt Times LT Std by Integra Software Services Pvt. Ltd, Pondicherry, India

CONTENTS

PREFACE

The lack of knowledge of steam system design and operation in the engineering community is becoming more evident each year. Consequently, most organizations tasked with designing or operating a steam system rely heavily on the steam equipment manufactures to help them figure it out. Although there is a lot of good individual equipment guidance available today, if you don't understand the steam system function as a whole, it is very easy to end up with a poorly designed system.

The book is written with a system approach and looks at each part of the steam system individually with many examples of how to size and select the right equipment for a variety of applications. In addition to engineering principles, the text discusses good operations, maintenance, troubleshooting, and commissioning practices of the entire steam system. This book is meant to provide a text for learning and a reference guide for operators, maintainers, and engineers.

WHAT'S NEW IN THE SECOND EDITION

The first edition proved to be a relatively simple but comprehensive collection of steam design and operational good practices. In this edition I add more design guidance, discuss more applications, include the historical importance of steam, show the development of boiler safety code, and provide a series of technical problems. In addition more discussion on controls and concepts of sustainable practices in the boiler room. The main focus remains on good engineering and operational practices with emphasis on what happens inside the steam system equipment at any location in the steam cycle. In this edition I rearranged some chapters to make better information flow.

HOW BEST TO USE THIS GUIDANCE

How you choose to use this material depends on what you intend to do with the information. Use as teaching material, an instructor can start from the beginning of the text and present the material in order as it's written. Each subsystem is presented with its own chapter. The problems in the back of the book will help readers apply theoretical concepts to solve practical system problems.

Maintainers and operators will find the troubleshooting section particularly useful as a library of potential solutions to steam system problems. The text will also provide them with the fundamental design principles so they can better understand why their system is designed the way it is or the way it should be.

Plant engineers, mechanical engineers, and process engineers will find the design basis information for all aspects of a well-designed steam system. There are numerous "rules of thumb" presented to help conceptualize a new design and perform a steam system assessment. In addition, detailed calculation methods are included to help seasoned engineers perform detailed designs. I have tried to explain the reasoning behind the design principles to help engineers explain these design criteria to plant leaders.

ACKNOWLEDGMENTS

This information contained in this book would have been much harder to assemble if it were not for some unselfish people. I thank the many good technical minds from folks like Paul Pack, Jim Pettiford, Jerry Porter, Jon Seigle, Lee Van Dixon, Tim Coughlin, and Doug Clark. Thanks to Eric Riehl for helping pull together some of the design detail illustrations. Laurie Iauco's administrative help putting some graphs and tables together was very useful. To all of these good folks, I am deeply grateful.

LIST OF EXAMPLES

LIST OF TABLES

1

STEAM: A HEAT TRANSFER FLUID

Steam provides a means of transporting controllable amounts of energy from a central boiler room, where it can be efficiently and economically generated, to the point of use. For many reasons, steam is one of the most widely used commodities for conveying heat energy. Its use is popular throughout industry for a broad range of tasks from mechanical power production to space heating and process applications. This is why some consider steam to be **the energy fluid** [1].

The ability of steam to retain a large amount of energy on a per weight basis (1000–1200 btu/lb) makes it ideal for use as an energy transport medium. Since most of the heat energy contained in steam is in the form of latent heat, large quantities of energy can be transferred efficiently at constant temperature, which is useful in many process heating applications.

The use of steam has come a long way from its traditional associations with nineteenth-century locomotives and the Industrial Revolution. Steam today is an integral and essential part of modern technology. Without the use of steam, our food, beverage, textile, chemical, medical, power, heating, and transportation industries would be crippled.

WHY STEAM?

Water can exist in the form of a solid (ice), a liquid (water), or a gas (steam). In this book, our attention will concentrate on the liquid and gas phases, and the equipment required to facilitate the change from one phase to the other. If heat energy is added to water, its temperature rises until a value is reached at which the water can no longer exist as a liquid. We call this the "saturation" point. Further addition of energy will cause some of the water to boil as steam. This evaporation requires relatively

Process Steam Systems: A Practical Guide for Operators, Maintainers, Designers, and Educators, Second Edition. Carey Merritt.

large amounts of energy, and while it is being added, the water and the steam are both at the same temperature. As the steam is formed in a closed vessel, it develops pressure allowing it to flow anywhere to a lower pressure, that is, through piping and distant equipment. Likewise, if we allow the steam to cool it will release the energy that was added to evaporate it. These boiling, transfer, and condensing events provide a simple mechanism to transfer energy from one place to another, hence the basis of a steam system. *Interestingly, steam is colorless; the white color often seen when steam discharges to the atmosphere is from the condensed water vapor in the steam.*

Steam Is Safe and Flexible

Water is plentiful and inexpensive. It is nonhazardous to health and environmentally sound. In its gaseous form, it is a safe and efficient energy carrier. **Steam can hold five to six times as much energy as an equivalent mass of water**. It can be generated at high pressures to give high steam temperatures. The higher the pressure, the higher the temperature, so it's potential to do work is greater. Modern boilers are compact and efficient in their design, using multiple passes and efficient burner technology to transfer a very high proportion of the energy contained in the fuel to the water, with minimum emissions. The boiler fuel may be chosen from a variety of options, including, natural gas, LP gas, oil, solid fuels, alternative fuels, and electricity, which makes the steam boiler an economical and environmentally sound option amongst the choices available for providing heat energy. Highly effective heat recovery systems can significantly reduce exhaust gas and water discharge energy losses creating an overall efficiency of the steam system approaching 85 percent. Boiler plants can be centralized or installed at the point of use. Sizes range from a few pounds of steam to thousands of pounds of steam per hour. Steam is one of the most widely used media to convey heat over distances. Because steam flows in response to the pressure drop along the pipe line, expensive circulating pumps are not needed. Not only is steam an excellent carrier of heat, **it is also sterile** and thus popular for process use in the food, pharmaceutical, and health industries. Other industries within which steam is used range from huge petrochemical and bio fuel plants to small local laundries. Further uses include the production of paper, plastics, textiles, beverages, food, metal, and rubber. Steam is also used extensively for power generation, humidification, and space heating.

Steam is also **intrinsically safe** – it cannot cause sparks and presents no fire risk. Many chemical plants and refineries utilize steam fire-extinguishing systems. It is ideal for use as a heat transfer media in hazardous areas or explosive atmospheres.

Steam Is Easy to Control

Because of the direct relationship between the pressure and temperature of saturated steam, simply controlling the steam pressure one can control the temperature of the steam and the process material being heated. Furthermore, the total amount of energy input to a process stream is directly related to the steam mass flow heating that process. Modern steam controls like pressure reducing valves and flow control valves are designed to respond very rapidly to process inputs. Therefore, today steam pressure and flow can be precisely regulated to add heat energy to a process. Industrial processes that have tight heating tolerances are well suited for process steam use.

The **heat transfer properties of saturated steam are high** and the required heat transfer area is relatively small. This enables the use of compact heat transfer equipment, which reduces installation costs and takes up less space in the plant. Most steam controls are able to interface with modern networked instrumentation and control systems to allow centralized control, as the case of a Building/Energy Management System or Process Computers. With proper maintenance, a steam plant will last for many years, and many aspects of the system are easy to monitor on an automatic basis.

In contrast, hot water and hot oils have a lower potential to carry heat energy per pound. Consequently large amounts of water or oil must be pumped around the system to satisfy process or space heating requirements. However, hot water is popular for general space heating applications and for low-temperature processes (up to 200 F) where some temperature variations can be tolerated. Furthermore, thermal fluids, such as mineral oils, may be used where high temperatures (up to 700 F) are required, but where high steam pressure is undesirable. An example would include using hot oil as the heating source in certain chemical process reactors. Figure 1.1 shows that saturated and superheated steam can provide a heating range from 200 to >1000 F.

THE CONCEPT OF STEAM FORMATION: BOILING

Perhaps the best way of visualizing the formation of steam in a boiler is to compare what happens in a pot of water being heated on a kitchen stove. See Figure 1.2. Addition of heat to the water in the pot raises its temperature, until a point where the temperature reaches the boiling point but does not boil yet. Thus far the heat added by the stove burner is called sensible heat. **Sensible heat** is the energy added or removed from a substance that corresponds to a temperature change only. Water that enters a boiler and heated up to the boiling point temperature, without boiling occurring, is considered to have a gain in sensible heat only.

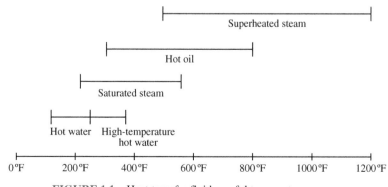

FIGURE 1.1   Heat transfer fluids useful temperature ranges.

FIGURE 1.2   A pot of boiling water: The simplest boiler. Markus Schweiss / Wikimedia Commons / CC BY-SA 3.0.

To allow the hot water molecules to change from the liquid state to a gaseous state, a large amount of additional energy must be added. This amount of heat energy is called the latent heat of vaporization. **Latent heat** is the heat added to a substance that does not cause a change in temperature, rather, creates a change in state. Latent heat is added when ice is changed to water and again when water is changed to steam. Figure 1.3 shows the relationship between temperature and energy content of water and steam [2, 3]. The curve shows two forms of heat gain, latent and sensible heat. Latent heat tends to be much greater than sensible heat. For instance heating water in a pot, open to the atmosphere, from 33 to 212 F requires only 179 btu, but changing the 212 F water to 212 F steam requires about 971 btu. The latent heat value is important because it not only is the amount of energy required to change water to a gas (steam), but is the same amount of energy steam will yield when it condenses in process equipment (i.e., heat exchanger). Notice, as sensible heat is added to steam above the **Saturation Temperature** (boiling point), the steam becomes superheated steam.

Pressure and Boiling

The pot of water on the stove mentioned above is subject to "atmospheric pressure."

This is simply the pressure exerted on all things, in all directions by the earth's atmosphere. The pressure exerted by the atmosphere at sea level, happens to be 14.7 psi (pounds per square inch) or 1 Bar. Water subject to atmospheric pressure will

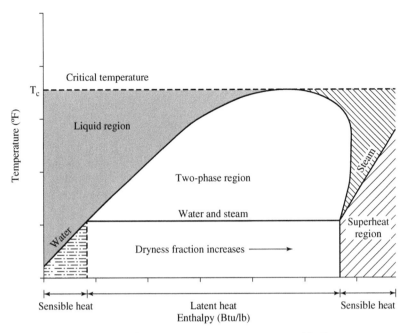

FIGURE 1.3 Water and steam enthalpy curve [2, 3].

boil at 212 F. The boiling of water in a pot on the stove is quite easy to visualize; however, we must consider the effects of pressure to really understand steam formation in a boiler. If we placed a pressure gauge in the pot of water on the stove, open to atmosphere, it would read 0 **pounds per square inch gauge or psig.** Therefore, any pressure shown on a pressure gauge mounted in the pressure vessel of a boiler, is in addition to atmospheric pressure. The sum of the gauge pressure and atmospheric pressure is called the **absolute pressure, psia or pounds per square inch absolute.** This relationship between gauge pressure and absolute pressure is important to recognize when reading steam charts. The relationship of absolute and gauge pressure is shown below.

Absolute pressure = gauge pressure + atmospheric pressure (14.7 psi at sea level)

It is to be expected that if the pressure above the liquid were increased, the molecules would need more energy to leave the liquid phase. This means the temperature of the water would increase before boiling occurred. Indeed, this is exactly what is found in practice. If our pot was fitted with a sealable lid allowing the pot to build pressure, then the temperature of the water could be increased above 212 F before any boiling commenced. Similarly, increasing the pressure in the boiler, will increase the temperature of steam the boiler produces, and the temperature rise in any substance heated up by that steam. By simply regulating steam pressure, you can effectively control the temperature of anything that is being heated with that steam.

Furthermore, as you rise in altitude above sea level, the atmospheric pressure goes down and the corresponding temperature that water will boil is lower. Equally, if the pressure of water is lowered below the normal atmospheric pressure (i.e., vacuum), then it is easier for the molecules to break free and the temperature at which boiling occurs, is lower than 212 F. This is the concept used to vacuum distill a process fluid.

The Ideal Gas Law

The ideal gas law states that a gas pressure, temperature, and volume are all related by the formula $PV = nRT$, where P is pressure, T is temperature, V is volume, n is the number of moles of gas, and R is a constant. It can be seen that if we do not change the moles of gas then PV/T = constant, and therefore

$$\frac{P_1 V_1}{T_1} = \frac{P_2 V_2}{T_2}$$

where the 1 and 2 subscript represent the before and after conditions. When the volume is held constant, pressure and temperature are directly related. Likewise, if we hold the pressure constant, the volume and temperature are directly related. Meaning as one goes up, the other must also go up. **This also shows us that pressure and volume are inversely related, consequently as one increases, the other must decrease**. Thus, as we increase steam pressure, the temperature of that steam will go up or the volume will go down. Using this formula, one can calculate any one of the three variables ($P_2, T_2,$ or V_2), if the other five are known. Fortunately for all of us, engineers and physicists have already carefully measured and recorded the temperature, volume, and energy amounts of steam at various pressures. Their results appear in published *Steam Tables* and also found in Appendix A, page A-3. Table 1.1 shows some example data from those charts. The relationship of total energy, sensible heat, latent heat-specific volume, and temperature is also shown in this data.

When water is converted to steam its volume may increase by a factor of over 1000 times. The actual volume increase during this conversion process is dependent on the pressure at which boiling occurs.

TABLE 1.1   Properties of Saturated Steam

Gauge Pressure (psig)	Temperature (F)	Sensible Heat (btu/lb)	Latent Heat (btu/lb)	Total Heat (btu/lb)	Specific Volume (ft³/lb)
0	212.0	180.2	970.6	1150.8	25.8
10	239.4	207.9	952.9	1160.8	16.5
50	297.7	267.4	912.2	1179.6	6.68
100	337.9	309.0	881.6	1190.6	3.90

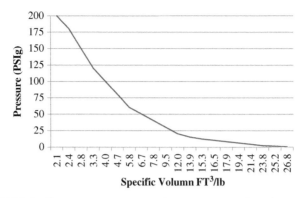

FIGURE 1.4 Steam pressure versus volume curve. Data from Steam Charts.

Example 1.1

A pint of water equals about 0.0167 ft³ but equals 25.8 ft³ of steam at atmospheric pressure, and only 3.9 ft³ at 100 psig.

This pressure versus volume relationship can be seen by looking at the data as found in the steam charts. Typically the volume of steam is listed as the **specific volume** of steam and is shown as cubic feet per pound of steam. The specific volume of steam versus pressure is shown in Figure 1.4 below. Notice the rapid increase in specific volume as the pressure approaches atmospheric pressure.

The specific volume–pressure relationship is a very important concept engineers must understand to adequately size steam piping. **The lower the steam pressure, the larger the volume the steam occupies for the same amount of mass flow and consequently, the larger the steam piping.** Likewise, steam boiler nozzles are sized to keep steam velocities below 5000 ft/s at some design pressure. Operating a boiler at a lower pressure than design can create very high steam velocities which will have a dramatic negative impact on steam quality. Quantifying this negative impact has been illustrated in this book in chapter 5 on Steam Quality.

2

THE DEVELOPMENT OF BOILER SAFETY

"If you have no desire to read this chapter, then please heed this. Never, ever add water to a running boiler being dry fired, otherwise it might be the last thing you do."

–The author who has seen the results of a boiler explosion

Those who operate or maintain a steam system should have an appreciation of the evolution of steam boiler design and the significance of boiler pressure vessel code. Steam played a critical role in western civilization's Industrial Revolution. Being able to generate an energy fluid and convert it into mechanical energy allowed for commodity mass production that we still see in today's economy. That fast-moving industrial progress came at a price. Boiler explosions have killed or injured hundreds of people, especially in the early years. This chapter will outline the development of steam use and show the lack of good operating and maintenance of the pressure vessel can lead to catastrophe. Back in the 1800s there were no computers to run stress analyses. Pressure vessels were riveted. Eventually, the engineering community adopted ASME codes that would prove to significantly improve the safety in the boiler room.

HOW IT ALL BEGAN

The history of steam use dates back centuries ago but really did not become useful as an industrial energy source until the early 1700s when Thomas Savery and Thomas Newcomen developed a steam engine used to extract water from mines in Europe. This invention made coal mining much more efficient. Water could be removed from a mine allowing for deeper shafts and access to more coal. The next breakthrough came around 1769 when James Watt improved the Newcomen engine by developing

Process Steam Systems: A Practical Guide for Operators, Maintainers, Designers, and Educators,
Second Edition. Carey Merritt.
© 2023 John Wiley & Sons, Inc. Published 2023 by John Wiley & Sons, Inc.

a steam motor that turned an axle instead of pumping water. Watt discovered the value of the condenser as part of the steam cycle. He learned that using a cooling loop to condense unused steam would create a vacuum and increase the engine's efficiency. Soon afterward he developed the governor, a concept of regulating steam flow and thus engine output. Watt's little 10 hp steam engine fueled by coal or wood was used to power a variety of manufacturing machinery. By the early 1800s Watt's steam engine could be used to drive ships that were being powered by coal-fired steam boilers. The first large commercial steam-powered ship to cross the Atlantic Ocean occurred in 1819 [4].

In 1829, English engineer Robert Stephenson used steam power to power a locomotive called his "Rocket." The concept quickly became a commercial success. By the late 1800s shipment by boats and trains connected the world markets. What really kicked the Industrial Revolution into high gear was the ability to generate electrical power using steam. A big step forward came in 1882 when George Babcock and Stephen Wilcox discovered the water tube boiler, and by the end of the 1800s the first steam turbine generators were developed to convert coal and biomass into electrical and mechanical energy. By the early 1900s commercial power plants using steam turbines were commonplace. The next century was spent developing more efficient and safer boiler designs, and discovering more applications to use steam.

THE CONSEQUENCES OF DEVELOPMENT

All this development came with a price. A boiler is a pressure vessel and each one is rated to contain its contents at a design pressure. Today, we see this stamped on the vessel name plate as MAWP or Maximum Allowable Working Pressure. In the early years pressure vessels were riveted by overlapping the metal pressure vessel walls and water level and pressure controls were archaic. Water chemistry impacts and corrosion mechanisms coupled with imperfect metal production technology resulted in much higher potential of a pressure vessel failing over time. This resulted in hundreds of deaths due to pressure vessel explosions. Boiler explosions were commonplace during the early years of boiler development. Figure 2.1 below shows the deaths from boiler explosions for every 15 years starting in the early 1800s.

Modern-day boiler explosions can occur via different scenarios. The most common is the result of water rapidly flashing to steam creating a massive pressure wave. The volume water will expand about 1600 times when flashed to steam. When a pressure vessel runs out of water and the burner continues to fire we have a condition known as "Dry-Firing." A dry fire condition can result in a furnace metal temperature over 2000 F. If water is introduced suddenly to the very hot metal, it is flashed to steam and the pressure vessel becomes a bomb. Figure 2.2 below shows the results of a dry-fired boiler explosion and a rapid opened steam valve. If an operator arrives in the boiler room and witnesses a dry fire event, turn the fuel supply off and go have a cup of coffee. ***Never attempt to add water to a dry-fired boiler, which is counterintuitive to what most people would do if they found a running boiler with no visible water level in the sight glass***. This is why all steam boilers come with two low water

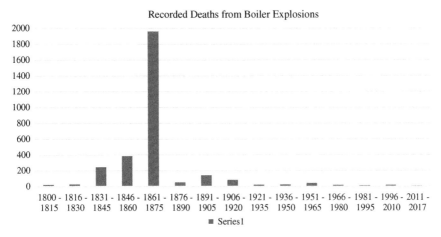

FIGURE 2.1   History of deaths from boiler explosions [5].

FIGURE 2.2   Results from an early locomotive boiler explosion that killed three men [6, 55] Unknown author / Wikimedia Commons / Public Domain.

cutoff controls. Unless these controls are jumpered out, a steam boiler should not run when the water levels falls below a preset level.

A second scenario that may result in a boiler explosion is the rapid opening of a steam valve allowing a massive rush of steam from the pressure vessel. This condition can force a slug of boiler water to instantaneously be carried into the steam line and block the steam outlet during a time when the boiler pressure is reduced. The resulting lower boiler pressure creates flash steam in the boiler is trapped in the vessel, creating a sudden pressure spike. The pressure spike can cause an immediate vessel failure or provide motive force to the slug of boiler water in the steam piping causing a water hammer consequence. Good steam system design that controls the steam flow from the boiler is the best way to prevent this condition.

The last scenario sometimes seen is a firebox or burner fuel explosion. This is a condition where fuel is added to the burner area after combustion is terminated. The build up of unused fuel can get to a level where ignition occurs. Gas and oil burners are susceptible to this type of explosion. Anyone who has operated an oil-fired furnace or boiler likely has noticed a drumming sound. This is a thumping sound which is generally caused by the rapid series of fuel detonations. If the air to fuel ratios are not correct, this condition can occur. Burners have fans with preset pre- and post-purge controls to void the combustion chamber of any unburnt fuel before the ignition sequence is allowed to commence. Burner controls incorporate flame detection safeties that automatically shut the fuel supply off if the flame extinguishes.

DEVELOPMENT OF ASME CODE

Through the early years deaths from boiler explosions were relatively commonplace. However, it wasn't until a major boiler exploded on a riverboat in 1865 that killed 1238 passengers, that the public demanded action be taken to boilers and pressure vessel safer [5]. The outcome was the formation of the Hartford Steam Boiler Inspection and Insurance Co. in 1866. Consequently, in 1879 Harford published the first boiler construction standard and in 1880 the American Society of Mechanical Engineers (ASME) was formed. In 1887 ASME issued its *Standard for the Diameter and Overall Dimensions of Pipe and Its Threaded Ends* in hopes of eliminating variations in pipe manufacturing.

As boiler explosions continued to occur the engineering community realized a greater need to protect the public. One of the more notable explosions occurred in 1905 at the Grover Shoe factory in Brockton Massachusetts in 1905. That event resulted in 58 deaths, 117 injuries, and completely leveled the factory. By 1915 ASME published the first *American Boiler and Pressure Vessel Code or BP&PVC*, which soon gained worldwide acceptance. The BP&PVC code grew to be the largest standard issued by ASME Codes and Standards [8]. It establishes safety rules governing the design fabrication and inspection during construction of boilers and pressure vessels.

The new ASME code provided guidance on how to ensure the structural integrity and mechanical safety of a boiler; it did not provide guidance on how to operate them. By 1924, The National Board of Fire Underwriters developed a document called NBFU 60 to address some of the operating protection concerns. This document eventually became an NFPA standard in 1946. We know it today as NFPA 85. This guidance required many safeguards like trips, interlocks, and permissives between specific control functions (i.e., water level and burner operation) to be included in the operational scheme of a boiler. This led to the development of burner management systems. It took almost 100 years and many deaths to develop the safety standards we use today. These regulations get reviewed and updated every three years to ensure progress continues.

FUTURE OF STEAM

The age of steam-driven ships and locomotives is long gone, and many low-pressure steam heating systems have been converted, rightly so, to higher efficiency hot water heating systems. However, the use of steam in the process and power generation industries remains strong and will likely continue that way well in to the future.

The use of steam will likely shift along with the energy market. Steam use for natural gas fracking and cleaning up the produced waters has increased recently. The polymer solutions injected into the fracking bore are sometimes heated with steam to reduce viscosity of the penetrating solution. Furthermore, for shallow oil wells, injecting steam is becoming an increasingly popular method of extracting heavy crude oil. The steam is injected directly into the subsurface reservoir to reduce the crude's viscosity and allows for easier removal. The large oil sands in Alberta Canada is one area that uses this technic, called "Cyclic Steam Stimulation." See Figure 2.4.

Likewise, gasification and pyrolysis of biomass, plastics, and municipal waste could play an important role as future energy sources. The gasification process uses steam in the reforming step to produce syngas. The syngas can be used to make a number of useful chemicals and fuel blends. Pyrolysis is the process of heating a carbon-containing material in the absence of oxygen. The net result is a bio oil and char. The bio oil can be further processed into a useful fuel. An important step in this process is recovering the waste heat in the form of steam generation. Consequently, Heat Recovery Steam Generators (or HRSGs) are used to generate steam used for electrical power production. One such commercially available system is shown in

BOILER EXPLOSION AT BEAVER MILLS, KEENE, N. H., MAY 22, 1893.

FIGURE 2.3 Results from a boiler explosion that occurred at the Beaver Mill Plant in 1893 [7] that killed three men. [43] Keene Public Library / Wikimedia Commons / Public Domain.

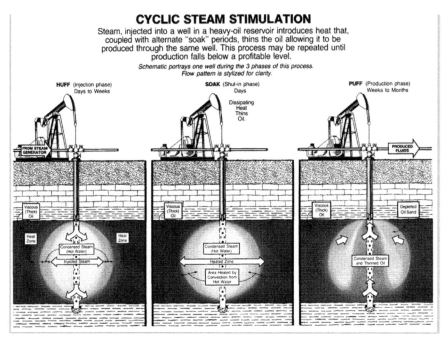

FIGURE 2.4   Cyclic steam stimulation [9] / U.S Department of Energy / Public Domain.

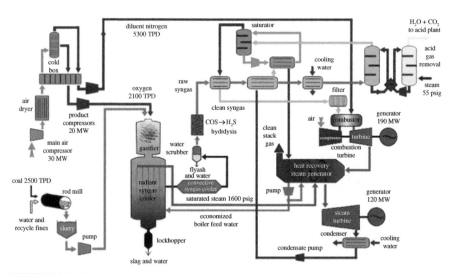

FIGURE 2.5   Biomass gasification process using an HRSG [10] / Stan Zurek / CC BY SA 3.0.

Figure 2.5. Considering the massive amount of waste material that is available and could be converted to a useful fuel via gasification or pyrolysis, steam use could play an important role in our future energy needs.

3

UNDERSTANDING HEAT TRANSFER

It is nearly impossible to fully grasp how a steam system works without a good understanding of heat transfer. A boiler in very simple terms converts fuel energy into steam. It does this by combustion and heat transfer. Overall system efficiency is directly related to heat transfer within the system. This chapter will show the reader what types of heat transfer take place within the steam system equipment and the importance of each. The reader should review this entire section before trying to apply any one heat transfer-type calculations as system heat balances should take into consideration the interrelationship of the three types of heat transfer.

Once combustion takes place in the boiler furnace, the thermal energy must be transferred to the water to make steam. Similarly, the energy in the steam must be transferred to a product to complete the conversion of fuel energy to product energy. We know that whenever a temperature gradient exists, transfer of heat energy will occur. This may take the form of either **conduction-, convection-, or radiation-type heat transfer**. The three types are shown in Figure 3.1. To be able to fully appreciate the efficiency of a steam system we must understand where, when, and what affects these three types of heat transfer mechanisms. The efficiency of all steam systems is optimized by optimizing desirable heat transfer and preventing unwanted heat transfer. The Figure 3.2 below shows where heat energy transfer occurs in a typical steam system.

RADIATION-TYPE HEAT TRANSFER

The heat transfer due to the emission of energy from surfaces in the form of electromagnetic waves is known as thermal radiation. One can sense this type of radiation heat transfer by simply putting your hand next to the hot metal surface like

Process Steam Systems: A Practical Guide for Operators, Maintainers, Designers, and Educators,
Second Edition. Carey Merritt.
© 2023 John Wiley & Sons, Inc. Published 2023 by John Wiley & Sons, Inc.

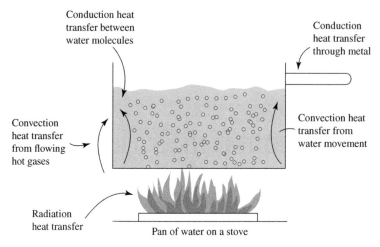

FIGURE 3.1   The three types of heat transfer.

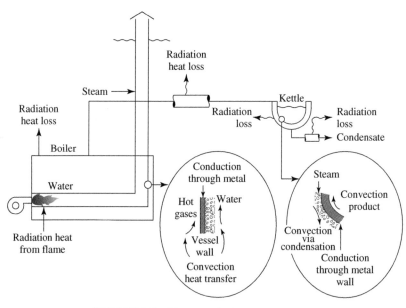

FIGURE 3.2   Heat transfer in a steam system.

a wood burning stove or an open flame. **Radiation heat transfer depends to a large extent upon the emissivity of the hot surface and the temperature between the hot surface and the colder surroundings**. Generally, smooth shiny surfaces are poorer emitters and absorbers of radiation than rough and dark surfaces. Radiation heat transfer is not dependent on material thickness, but is dependent on surface temperature. Table 3.1 shows some emissivity's of some common surfaces.

TABLE 3.1   **Emissivity Coefficient (ε) for Some Common Materials**

Surface Material	(ε) Emissivity Coefficient
Aluminum sheet	0.09
Brass polished	0.03
Cast iron	0.65
Glass, Pyrex	0.90
Ice, smooth	0.96
Iron plate, gray polished	0.31
Iron plate, red rusted	0.61
Porcelain, glazed	0.92
Brick	0.93
Steel, oxidized	0.79
Stainless steel, weathered	0.85
Stainless steel, polished	0.07
Water	0.95
Wood, planed	0.90

Data from *North American Combustion Handbook* [11].

The radiation energy given off to its surroundings per unit time is expressed as [12]

$$q = \varepsilon \sigma \left(T_h^4 - T_c^4 \right) A$$

where

q = net radiation heat loss rate (btu)
ε = emissivity coefficient
$\sigma = 0.1714 \times 10^{-8}$ btu/h ft^{2}R^4, The Stefan Boltzmann Constant
T_h^4 = hot body temperature in degrees R
T_c^4 = cold surroundings temperature in degrees R
A = area of the hot body (ft^2)

It should be noted that radiation heat is proportional to the fourth power of the surface temperature. Radiation heat transfer occurs in two main areas is a steam system; internally in the furnace area and externally anywhere there is a hot surface exposed to cooler ambient air. The burning flame (>1500 F) in the boiler burner will create radiation-type heat that is absorbed by the boiler pressure vessel. This section of a boiler furnace generally referred to as the "first pass" where the burner flame is visible. A significant portion of the combustion flue gas energy can be transferred to the pressure vessel via radiation-type heat transfer. Fuels that burn with higher flame temperatures, can create more radiation-type heat transfer. Likewise, burner design and boiler furnace configuration (area) will also affect radiation heat transfer. Good radiation heat transfer is desirable and is critical for boiler efficiency.

Undesirable radiation heat transfer can also occur at any hot surface in the steam system. The boiler itself will radiate heat energy. Uninsulated manways, water columns, steam and feed water piping will radiate heat and contribute to heat loading in the boiler room. The same can be said for any exposed hot tank surfaces or valve bodies in the steam system. Insulation reduces surface temperature and radiation heat losses, and should be employed wherever practical.

Example 3.1
A boiler room houses two steam boilers with a total surface area of 1000 ft² and operate with a skin temperature of about 120 F. The maintenance department asks if they could shut off the room ventilation for 4 h for some PM work. You are tasked with determining what heat load the boilers will contribute to the room with the ventilation off. The room temperature will be 90 F when the work starts.

Answer: Since there will be no air circulation, then you could assume the heat given off from the boilers will mainly be from radiation. The data suggests from the proceeding formula [12] A = 1000 ft², T_h is 120 F = 579.6R, T_c is 90 F = 549.7R, and ε = 0.79 from Table 3.1. Then q = 0.79 × 0.1714 × 10⁻⁸ btu/h ft²R⁴ × (579.6R⁴ – 549.6.7R⁴) × 1000 ft² = 29,175 btu/h. This data used in conjunction with air volume and humidity level in the room can be used to determine the air temperature rise in the 4-h period.

Radiation-type heat transfer is also greatly augmented by convection. As heat energy is emitted from a hot surface, the local ambient temperature rises, the delta temperature and radiation heat transfer starts to diminish. Convection can sweep away the ambient air and replace it with cooler air. More on this collaborative affect later in this chapter under convection-type heat transfer.

CONDUCTION-TYPE HEAT TRANSFER

When a temperature gradient exists within either a solid or stationary fluid medium, the heat transfer which takes place is known as conduction. This phenomenon can be seen in solids, liquids, and gases. In solids, the molecules are closer together relative to liquids and conduction will occur much faster compared to liquids. Similarly, molecules in liquids are much closer than gases and conduction in liquids is greater than in gases.

Conductance is often referred to as thermal conductivity. **Thermal conductivity (k) is defined as the quantity of heat (Q) transmitted through a unit thickness (L) in a direction normal to a surface of unit area (A) due to a temperature gradient (ΔT) under steady-state conditions and when the heat transfer is dependent only on the temperature gradient** [11]. The equation below shows how thermal conductivity is calculated and has units of btu/(h-F-ft) or W/(m-K).

$$k = Q \times L / \left(A \times \Delta T \right)$$

The Table 3.2 below shows the variation of thermal conductivity with temperature for various common materials.

TABLE 3.2   **Thermal Conductivities for Some Common Materials [11, 12]**

Material	(k) Thermal Conductivity in btu/(h-ft²) (°F/ft Thickness)
Low carbon steel	31 @ 77 F and 27 @ 437 F
Stainless steel, 304,316	9.4 @ 212 F and 12.4 @ 932 F
Cast iron	27–46 @ 70 F
Aluminum	145 @ 77 F and 145 @ 437 F
Silver	242 @ 77 F
Admiralty brass	64 @ 68 F
Water	0.34 @ 68 F and 0.39 @ 300 F
Air	0.02 @ 77 F
Dry steam	0.03 @ 257 F
Fiberglass insulation	0.04 @ 68 F
Glass	0.2 @ 68 F
Waterside scale	1.3 @ 400 F
Carbon black (soot)	0.144 @ 133 F
Wood	0.1 @ 68 F
Sawdust	0.034 @ 68 F
Slag, blast furnace	0.064 @ 68 F
Ice	1.3 @ 32 F
Mineral wool	0.02 @ 68 F
Snow	0.27 @ 32 F
Ethylene glycol	0.15 @ 32 F
Flue gases	0.26 @ 500 F
Oils	0.08 @ 86 F

Air has a particularly low thermal conductivity and this is why insulating materials often have lots of air spaces. In addition to air, water, scale, dry steam, and glass all have relatively low thermal conductivities. It is worth noting that not all metals have equal thermal conductivities. This is important to consider for heat exchanger design and selection.

Another way to consider thermal conductivity is to realize it is the inverse of resistivity (R-factor) multiplied by the material thickness. Therefore, thermal conductivity = d/thermal resistivity or $k = d \times 1/R$, where d is the material thickness and R is the resistivity. In a steam system, conduction heat transfer occurs mainly through the boiler pressure vessel and the steam/product wall boundary. Conduction also occurs to a lesser extent through the hot flue gases in the furnace area, the boiler water and through insulation of hot surfaces. Like radiation heat transfer, conduction is augmented by convection heat transfer.

The first area where conduction takes place is in the boiler furnace area. In addition to radiant heat energy transfer from the burner flame, the hot burner flue gases will conduct heat energy through the gas molecules to the boiler pressure vessel wall as the flue gases travel through the boiler furnace section. When the hot flue gas molecules contact the metal boiler pressure vessel, their energy is transferred quickly through the vessel wall to either the boiler water or to an insulation barrier. **Because thermal**

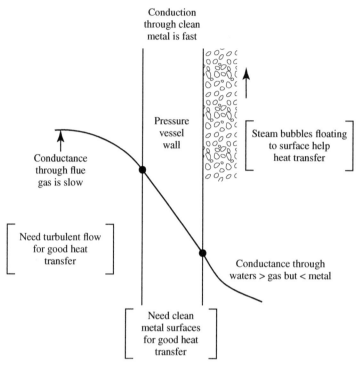

FIGURE 3.3   Relative thermal conductance through flue gas, pressure vessel, and boiler water.

conductance in gases is low, turbulent gas flow through entire the boiler furnace section is desirable to increase the probability of a hot gas molecule contacting the pressure vessel wall and surrendering its energy to the boiler water. This is the design basis for flus gas turbulators or specially designed boiler tubes that create flue gas turbulence. The heat transfer resulting from turbulent flue gas also creates convection-type heat transfer and discussed in detail below. See Figure 3.3

Conductance through the metal pressure vessel wall occurs quickly, however, if metal surface gets coated with any material that has a lower thermal conductivity, the overall thermal conductivity will become lower. Soot residue from incomplete combustion, hardness scale, or corrosion product sludge all have lower thermal conductivities than clean metal vessel wall. Therefore, when a boiler pressure vessel becomes sooted up, scaled or fouled, the thermal conductivity is lower thus reducing heat transfer efficiency. Energy managers should watch boiler pressure vessel cleanliness carefully as part of their boiler plant efficiency plan. The diagram below in Figure 3.4 shows the effect of foreign material on conductance through a metal wall. The scale and soot layer will reduce the overall conductive heat transfer through the metal wall.

Conduction also occurs in the boiler water. The furnace pressure vessel wall transfers heat energy to the boiler water at the waterside of the vessel wall. This heat energy in the film of water is then transferred to adjacent lower-energy water molecules via conductance. The rate of this transfer is less than the rate of transfer through the metal wall so

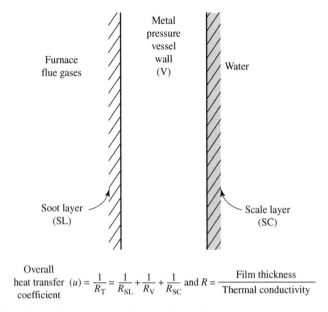

Overall heat transfer coefficient $(u) = \dfrac{1}{R_T} = \dfrac{1}{R_{SL}} + \dfrac{1}{R_V} + \dfrac{1}{R_{SC}}$ and $R = \dfrac{\text{Film thickness}}{\text{Thermal conductivity}}$

FIGURE 3.4   Heat transfer through metal wall with films attached.

convection is required or the metal will build heat energy. Fortunately, in a steam boiler, the high-temperature water and steam bubbles formed at the waterside wall are less dense and will rise to the top and create enough turbulence to remove enough heat to prevent the metal pressure vessel wall from overheating. Conduction also occurs in the equipment that condenses the steam. Heat transfer though the product in done via conduction through a heat exchanger wall and within the product itself. It is worth noting that dry steam or superheated steam has low conductance. This is one reason why using only superheated steam in a heat exchanger will generally require more surface area to transfer the same amount of energy relative to saturated steam. The minute amounts of entrained water in saturated steam increase the conductance.

Lastly, undesirable conduction occurs through the insulation of any hot metal wall. The entrained air spaces in the insulation will retard the conductive heat transfer and minimize energy loss. It is for this reason that the most effective insulating materials are made up of a mass of minute air cells, enclosed by nonconducting fibers. Using thermal conductivity data, for a fixed surface area a layer of air only 0.05 inches thick can offer the same resistance to the flow of heat as a layer of water 1 inch thick, or a layer of iron 4.5 feet thick.

CONVECTION-TYPE HEAT TRANSFER

The transfer of heat energy between a surface and a moving fluid of different temperature is known as convection. Convective heat transfer may take the form of either forced convection or natural convection. Forced convection occurs when fluid flow is induced by an external force, such as a pump or an agitator. Conversely, natural convection is caused by buoyancy forces, due to the density differences arising from the temperature variations in the fluid. The transfer of heat energy caused by a phase change, such as boiling or condensing, is also referred to as a convective heat transfer process. The equation for convective heat transfer is [11]:

$$Q = h_c \times A \times \Delta T$$

where

Q = the amount of heat transferred per unit time (btu)
A = the surface area (ft^2)
h_c = the convective heat transfer coefficient of the process (btu/ft^2-$^\circ$F)
T = the temperature difference between the surface and the bulk fluid (F)
And h_c for some common fluids is shown in Table 3.3

Surface flow has the largest influence on convection heat transfer. The effect of air flow over hot uninsulated surfaces can be seen in the data in Table 3.4.

Convection occurs on both the furnace side and the waterside of the boiler pressure vessel. On the furnace side, turbulent combustion gases scrape the pressure vessel

TABLE 3.3   Convective Heat Transfer Coefficients for Some Common Fluids

Fluid	Type of Convection	h_c in (btu/ft^2-h-F)
Air	Natural	1–5
Air	Forced	5–50
Oil	Forced	10–300
Water	Forced	50–2000
Water	Boiling	500–5000
Steam	Condensing	1000–10,000

Data from *North American Combustion Handbook* [11].

TABLE 3.4   Effect of Airflow Versus Heat Loss over Hot Steel Surfaces @ 80°F Ambient Air Temperature

Steel Temp (°F)	Radiation Heat Loss Only (btu/ft²-h)	Radiation and Convection Heat Loss (btu/ft²-h)		
		Horizontal Cylinder Natural Convection	Vertical Cylinder or Flat Surface Natural Convection	Vertical Cylinder or Flat Surfaces with 10 ft/s Forced Convection
100	22	25	25	
150	87	125	137	312
200	170	240	265	565
250	275	385	430	845
350	561	760	840	1455
500	1242	1625	1750	—
1000	7250	8210	8600	—

Data from *North American Combustion Handbook* [11].

wall and transfer energy to the wall metal. Since the thermal conductivity of the combusted fuel gases is low, turbulent flow is required to have convection type heat transfer in the boiler furnace to have any hope of heat transfer efficiency. **The consequence of high turndown boiler burners may create laminar combustion gas flow in the furnace and substantially reduce convective heat transfer**.

Convection also occurs on the waterside of the pressure vessel. When the boiler water is heated up at the pressure vessel wall, the change in density creates a small amount of convection. Further heating creates steam bubbles that collapse or rise to the surface causing turbulence near the heat transfer surface. This natural convection is critical for efficient heat transfer. Likewise, a product contained in a vessel being heated by steam can reduce its heat-up time significantly by enhancing convective heat transfer. When steam condenses and gives up its energy to the vessel wall and eventually to the product, the more the product is moved along the wall, the better the heat transfer. This is why materials being heated by steam are almost always mixed via an agitator or some other method of forced circulation.

THE HEAT TRANSFER EQUATIONS

In most practical situations, it is very unusual for all energy to be transferred by one mode of heat transfer alone. The overall heat transfer process will usually be a combination of two or more different types of heat transfer. Two simple equations can be used to calculate heat transfer: the general heat transfer equation and the product heat gain/loss equation. The **general heat transfer equation** measures heat flow from one body to another via all types of heat transfer. It is calculated as follows [13]:

$$Q_m = m U \Delta T_m$$

where

Q_m = mean heat transferred per unit time (btu/h)
U = overall heat transfer coefficient (btu/h × sq. ft. × F)
m = heat transfer area (sq. ft.)
ΔT_m = mean temperature difference between the primary and secondary fluid (F)

The Overall Heat Transfer Coefficient (U)

This takes into account both conductive and convective heat transfer between two fluids separated by a solid wall. The overall heat transfer coefficient is the reciprocal of the overall resistance to heat transfer, which is the sum of the individual resistances.

An accurate calculation for the individual heat transfer coefficients is a complicated procedure, and in many cases it is not possible due to some of the parameters being unknown. Therefore, the use of established typical values of overall heat transfer coefficient will be suitable for practical purposes. From Figure 3.4, the (U) value is

$$U = 1 / R_T = 1 / R_1 + R_2 + R_3$$

where

R_1 = resistance of the soot film
R_2 = resistance of the metal wall
R_3 = resistance of the scale film
R_T = total resistance of the barrier

On the burner side of the furnace wall a film of soot, from a poorly adjusted burner, can cling to the metal wall and all act as a barrier to efficient heat transfer. Likewise, sludge or scale build up on the waterside of the pressure vessel will also inhibit heat transfer.

Some general film data is shown below. Regular cleaning, boiler tuning, and water treatment are necessary to minimize the layer of scale, soot or sludge. Boiler heat transfer efficiency loss due to resistance films [14] are shown below.

Waterside	Furnace Side
1/64″ scale – >1% loss	1/32″ soot – >2% loss
1/32″ scale – >2% loss	1/16″ soot – >4% loss
3/64″ scale – >3% loss	1/8″ soot – >8% loss
1/16″ scale – >4% loss	

Example 3.2
An older 200 hp oil-fired boiler with a rated fuel input of 8.6 MMbtu/h was removed from service and inspected. It was discovered that the furnace side of the pressure vessel had a 1/16″ layer of soot on about ½ the furnace wall area and had 1/64″ of scale on the entire waterside. The cost of cleaning the boiler is about $5000. What is the payback period for cleaning?

Answer: The payback period will dependent on how much the boiler is used and the cost of fuel. This boiler operates at an equivalent of 180 days/year at full rate. The annual fuel usage of 8.6 MMbtu/h (manufactures data sheet) would be about 180 days × 24 h per day × 8.6 MMbtu/h = 37.152 MMbtu per year. At $5/MMbtu fuel cost, the cost to operate this boiler (clean and new) will be $185,760 per year. The fouling on the furnace side will reduce the efficiency by at least 2% and the scaling will further reduce the efficiency by 1%. Therefore the fouled boiler will require $185,760/100 – 3% = $191,505 per year or $5745 more per year than a clean boiler. Payback period is 5000/5745 × 1 year = 0.87 years or a little over 10 months.

Table 3.5 shows some widely used heat transfer coefficients (U values). Notice there are ranges of values listed in the table that cover a variety of heat transfer applications.

On the product side of a heat exchanger wall, there may exist a stagnant film of product. The heat flow can be greatly reduced by the resistance of this film. Agitating the product in some way will reduce the thickness of the stagnant product film. The effect of a water film on the steam side of a heat exchanger is equally detrimental. We know that when steam comes into contact with the cooler heat transfer surface, it gives

TABLE 3.5 (U) Values in btu/ft²-h-F for Heat Transfer Applications

Air to air, heat exchanger	20
Air to water, unit cooler	4–10
Flue gas to air, air preheater	4–10
Flue gas to steam, superheating	2–15
Flue gas to water, economizer	2–10
Steam to air, heating coil	10
Steam to oil, tank heater coil	10–12
Steam to water, heat exchanger	200–1000
Water to oil, heat exchanger	20–60
Water to water, coil in vessel, no forced convection	20
Water to water, coil in vessel, forced convection	150–300

Data from *North American Combustion Handbook* [11].

up its latent heat and condenses. The condensation may produce droplets of water, or a complete film may be formed immediately. Even if drop wise condensation takes place, the drops will very often run together and form a film. The water film has a surprisingly high resistance to heat transfer. Even though the thermal conductivity of steam is lower than water, the convective heat transfer from the steam condensing will create more heat transfer than a film of water. A very thin film of water provides a significant obstruction. A film of water only 1/100″ in thickness offers the same resistance to heat transfer as a ½″ thick wall of iron or a 5″ wall of copper. This underlines the importance of providing a dry steam supply, and of ensuring rapid removal of the condensate from the steam space via good drip leg and steam trap selection.

Mean Temperature Difference (ΔT_M)

The temperature may not be linear across on all areas of a heat transfer surface. For instance, product being heated through a heat exchange will heat up in a nonlinear way as it passes through the exchanger as shown in Figure 3.5. Therefore we must calculate the mean temperature to enable accurate calculation of the overall heat transfer.

The determination of the mean temperature difference uses the primary fluid (steam) temperature as a constant. However, as the secondary fluid passes over the heat transfer surface, the highest rate of heat transfer occurs at the inlet and progressively decays along its travel to the outlet. The rise in secondary temperature is nonlinear and is best represented by a logarithmic calculation. For this purpose, the mean temperature difference chosen is termed the Logarithmic Mean Temperature Difference or LMTD or ΔT_M.

ΔT_M or LMTD can be expressed as $\Delta T_m = \left(\Delta t_2 - \Delta t_1\right) / \ln\left(\Delta t_2 / \Delta t_1\right)$ [15]

where

Δt_2 = the larger temperature difference between the two fluid steams at either the entrance or exit of the heat exchanger

Δt_1 = the smaller temperature difference between the two fluid streams at either entrance or exit of the heat exchanger

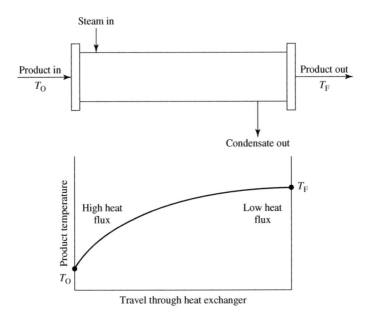

FIGURE 3.5   Temperature profile across a steam source heat exchanger.

ΔT_m for a Steam Boiler

A steam boiler is actually a pressure vessel that serves as a heat exchanger. In this application the flue gases are >1500 F in the furnace area, but cool to <500 F by the time they are exhausted from the boiler. The waterside of the heat exchanger is maintained at a constant temperature when the boiler is operating a steady state. The boiler water is considered the primary fluid and will be dependent on the boilers pressure. The secondary fluid is the combustion flue gas. To determine the LMTD or ΔT_M of a steam boiler we will need to know the flue gas temperature at the burner and at the boiler exhaust and, the boiler water temperature

Example 3.3

A steam boiler operating at 100 psig burns natural gas fuel and has a flue gas exit temperature of 450 F. What is the ΔT_m for this boiler?

Answer: For this boiler pressure we know from the steam charts that the primary fluid temperature (boiler water) will be 338 F. Combustion data shows the natural gas flame temperature is about 1800 F.

So $\Delta T_2 = (1800\ F - 338\ F) = 1462\ F$ and $\Delta T_1 = (450\ F - 338\ F) = 112\ F$

And $\Delta Tm = (1462\ F - 112\ F)/ln\ (1462\ F/112\ F) = 1350\ F/2.57 = $ **526 F**

Note: LMTD of 526 F is lower than the average temperature difference (1800 F + 450)/2 – 338 F = 787 F.

ΔT_m for a Steam to Process Fluid Heat Exchanger

It is often useful to size a heat exchanger that uses steam as the primary heating fluid. Since the steam condenses and only latent heat is transferred, the temperature of the primary fluid is constant and determined by the steam pressure. To determine the logarithmic mean temperature difference across the heat exchanger, we need the steam temperature, the product inlet and outlet temperatures.

Example 3.4

We have a plate and frame heat exchanger that uses 80 psig steam to heat up a liquid process stream from 120 to 300 F. What is the ΔT_m for liquid process stream?

*Answer: 80 psig steam is 324 F. The actual $\Delta T_2 = 324\ F - 120 = 204\ F$ and $\Delta T_1 = 324\ F - 300\ F = 24\ F$, and $\Delta T_m = 204\ F - 24\ F/\ln (204\ F/24\ F) = 180\ F/2.14 = $ **84 F**. 84 F is significantly different than the average temperature difference $(120\ F + 300\ F)/2 = $ **210 F**. Using the wrong ΔT when sizing a heat exchanger can create a poor design. Design engineers should always use the LMTD when determining the surface area required for a heat exchanger application.*

Surface Area (*m*)

The overall heat transfer is directly related to the square footage of heating surface. Higher surface areas result in more heat transfer. Consequently, boilers and steam using equipment with high surface areas tend to be more efficient because more there is more opportunity for heat to be transferred from the energy source to the energy recipient. Unfortunately, the same is true for unwanted heat transfer in a steam system. A 6-inch steam pipe will lose more heat per foot than a 4-inch steam pipe, due to the higher surface area of the larger pipe. The magnitude of the difference is related to the degree of insulation on either pipe.

Due to cost of manufacturing and space limitations, equipment manufacturers have minimized the surface area in modern day steam using equipment. To achieve maximum heat transfer from the hot flue gases to the boiler water, efficient use must be made of every square foot of the boiler heating surface. **Steam system designers should use steam boiler surface area as one criteria for selection**. Savvy marketing people will advertise their boiler as having "multiple passes." This refers to the number or times the flue gases change direction along or through the pressure vessel. A 2-, 3-, or 4-pass boiler is less important relative to the true surface area, which is the real measure of value.

Modern equipment with low surface areas can be more susceptible to conductive heat transfer efficiency loss from waterside films like scale and sludge. Consequently, current steam equipment requires high cleanliness to be efficient. If part of the heating surface is covered up with a film of lower thermal conductivity, the heat transfer that take place from the steam to the product will be reduced accordingly. This also happens in heat exchanger if the condensate is allowed to collect in the bottom of the steam space. A portion of the heating surface will be covered by water and will reduce the heat transfer from the steam to the product. This is commonly referred to as "water logging."

The second useful heat transfer equation is the **heat gain/loss equation**. The heat gain/loss equation can be used to determine the amount of heat energy gained or lost during a process. This is very useful in determining how much heat energy must be added or removed from a process. The heat gain/loss can be calculated as [13]:

$$Q = M \times \mathrm{Cp} \times \Delta T$$

where Q is in btu/h, M is the mass flow of material in lb, Cp is the material heat capacity or specific heat in btu/lb-F, and ΔT is the temperature rise or fall in F. For instance, when water is heated in a hot water tank in your home, the heat gain/loss equation can be used to determine the amount of energy needed to raise the temperature to a specified level. The heat gain/loss equation does not consider thermal conductivity of the conductive interface. It only accounts for the total heat energy transferred from a heat source to a product. If heat capacities are known, like the data shown in Table 3.6, and the mass is known, then simply measuring the temperature rise/fall will yield enough information to determine the amount of heat energy transferred to the substance.

TABLE 3.6   **Heat Capacities of Some Common Materials**

Material	Heat Capacity (btu/lb-°F)
Water	1.0
Aluminum	0.22
Concrete	0.19
Copper	0.09
Steel	0.12
Glass	0.20
Ice (@ –4 F)	0.47
Rubber	0.48
Wood	0.32–.48
Ethanol	0.60
Fuel oil	0.4–.5
Olive oil	0.47
Gasoline	0.53
Air, dry	0.17
Methane	0.45
Steam	0.35
Orange juice	0.43
Potatoes	0.41
Turkey	0.35

Data from Spirax Sarco [16].

Example 3.5

A 40 gallon hot water tank receives 50 F water and is required to heat the 40 gallons up to 125 F in one hour. How much energy is required to meet this requirement?

Answer: Using the heat gain/loss equation, $Q = M \times Cp \times \Delta T$, and we know M to be 40 gallons $\times$ 8.2 lb/gal = 328 lb, Cp to be 1 btu/lb-F, and the ΔT to be (125 – 50)F = 75 F. Solving for Q, we calculate the amount of heat energy to be **24,600 btu/h**.

Process engineers need to be able to make more complex calculation like to one below on a routine basis.

Example 3.6

We have a 100 sq. ft surface area plate and frame heat exchanger that uses 100 psig steam to heat up a 50 gpm liquid process stream from 120 to 275 F. The process flow rate and steam pressure are fixed; however, we would like to raise the process liquid temperature to 300 F. What can we do the resolve this problem? The process fluid has a density of 9 lb/gal. and a heat capacity of 0.9 btu/lb-F.

Answer: One solution to this problem is to retrofit this heat exchanger with more surface area (i.e., add more plates) or install a new larger heat exchanger. To determine the extra surface area needed, we can use the general heat transfer equation for each condition to solve for the new surface area. The heat transfer equation $Q = m \times U \times \Delta T$ can be solved for both conditions. ΔT_{m1} and ΔT_{m2} can be calculated as ΔT_{m1} is $\Delta T_2 = 338\,F - 120 = 218\,F$ and $\Delta T_1 = 338\,F - 275\,F = 63\,F$, and $\Delta T_{m1} = 218\,F - 63\,F/\ln(218\,F/63\,F) = 155\,F/1.24 = 125\,F$. Likewise, the desired ΔT_{m2} will be $218\,F - 38\,F/\ln(218\,F/38\,F) = 180\,F/1.75 = 103\,F$. Using this information the general heat transfer equations become $Q_1 = m_1 \times U_1 \times \Delta T_{m1}$, where $m_1 = 100\,ft^2$, $\Delta T_{m1} = 125\,F$, Q1 is unknown, U is constant, and $Q_2 = m_2 \times U_2 \times \Delta T\,m_2$, where m_2 and Q_2 are unknown, $\Delta T_{m2} = 103\,F$ and U is constant. We know that $U_1 = U_2$, so solving each equation for U and substituting one for the other, $m_2 = Q_2 \times 100\,ft^2 \times 125\,F/Q_1 \times 103\,F$, but we need to know Q_1 and Q_2.

*Using the heat gain/loss equation $Q = M \times Cp \times \Delta T$. Q_1 can be calculated by multiplying the process stream mass flow times the heat capacity $\times$ the ΔT or $Q = 50\,gal/min \times 9\,lb/gal \times 0.9\,btu/lb\text{-}F \times (275 - 120)F = 62,775\,btu/min$. And $Q_2 = 50\,gal/min \times 9\,lb/gal \times 0.9\,btu/lb\text{-}F \times (300 - 120)F = 72,900\,btu/min$. To find the relative increase in surface area, we can now substitute Q_1 and Q_2 to find m_2. So, $m_2 = Q_2 \times 100\,ft^2 \times 125\,F/Q_1 \times 103\,F$, where $Q_1 = 62,775\,btu/min$ and $Q_2 = 72,900\,btu/min$, so $m_2 = 72,900\,btu/min \times 100\,ft^2 \times 125\,F/62,775\,btu/min \times 103\,F =$ **141 ft^2**.*

HEAT FLUX

Heat flux is simply the amount of energy transferred through a unit area. It is generally expressed in btus per square inch or watts per square centimeter.

$$\text{Heat flux} = Q_{\text{m}}/m$$

Heat flux is a concept that is important, but often overlooked. The burner on a boiler converts chemical fuel energy to heat energy. This energy must be transferred through the pressure vessel wall to the boiler water to make steam. There is an increasing trend of manufactures fabricating boilers with smaller surface area pressure vessels. This saves space and fabrication costs, but increases the heat flux. High heat flux can put added thermal stresses on the boiler pressure vessel walls. In addition, scaling potential increases with increasing heat flux. Scaling is a consequence of decreasing calcium and magnesium salt solubility with increasing temperature. The higher heat flux will create increased potential for the calcium and magnesium salts to precipitate on the waterside of the boiler pressure vessel. Boilers with high heat flux must have good water treatment schemes or scaling will occur relatively quickly. The scaling affect is exacerbated by low boiler water volumes. Boilers with low water volumes and high heat flux need to be monitored closely. Low heat flux boilers are more forgiving than high heat flux boilers and should be used whenever possible. **A good rule of thumb is a steam boiler should have a minimum of 3 sq ft of heating surface area per boiler horsepower**.

4

STEAM FORMATION, ACCUMULATION, AND CONDENSATION

THE BOILING PROCESS

The boiling process is a rather simple process, but not fully appreciated by many. Inside a boiler, hot (>1500 F) combustion gases flow thru the furnace side of the pressure vessel. The pressure vessel boundary (i.e., wall) is the conduit by which the heat energy is transferred from hot combustion gases to the boiler water. The boiling process includes bubble formation, bubble rising, and bubble collapsing at the water surface.

Sounds simple…not quite. Steam quality and heat transfer are affected by this boiling process. As the hot gases transfer energy thru the boiler pressure vessel wall, the water gets heated up until enough energy is added to form a steam bubble at the waterside furnace wall. Much like the heated pot of water on the stove will form small bubbles near the bottom of the pot. The initial bubbles form and collapse as they give up their energy to the surrounding water. Eventually enough energy is added to liberate a bubble and boiling commences. In the pot of water, boiling occurs as **Nucleate Boiling**, or the continuous formation and liberation of bubbles. A traditional convection type boiler operated at steady state will have this desirable type of boiling. **During nucleate boiling the heat transfer through the pressure vessel wall is even and rapid, creating metal wall temperature very close to the water saturation temperature in the boiler** [17].

However, if heat is added at a very high rate to a boiler pressure vessel or the heat is not quickly transferred to the water, boiling may occur as **Film Boiling. The two types of boiling are shown in Figure 4.1**. Film boiling is found in high-pressure fossil fuel or electrode power boilers and in nuclear reactors. **Film boiling in a convection boiler can retard the transfer of heat energy to the water and create high-pressure vessel wall temperatures**. Consequently, this type of boiling in a

Process Steam Systems: A Practical Guide for Operators, Maintainers, Designers, and Educators, Second Edition. Carey Merritt.
© 2023 John Wiley & Sons, Inc. Published 2023 by John Wiley & Sons, Inc.

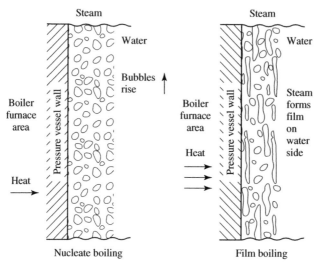

FIGURE 4.1   Nucleate and film boiling.

standard convection steam boiler is not desirable. The film of steam on the waterside pressure boundary will insulate the vessel and slow heat transfer to the water. Consequently, boilers designed for film boiling must has pressure vessel design considerations that account for this higher pressure vessel wall temperature gradient. Usually these types of boilers will require forced water circulation to minimize the film layer thickness and enhance heat transfer.

Steaming

Savvy marketing people in the boiler business use quick start up periods as a benefit of their design. Quick start-up times, however, can be a sign of a low mass, low water volume boiler, both characteristics that can be unfavorable in certain process applications. Obviously large metal mass and high water volume boilers require more heat energy input to get steam generation started, and consequently longer start up periods. These two boiler characteristics are important considerations in selecting the type of boiler for a given application.

One can determine what the actual start period is by considering what is happening in the boiler. The chemical energy contained in the boiler fuel is converted into heat energy when the fuel is combusted. This heat energy is transmitted through the wall of a boiler furnace (pressure vessel) to the water. Prior to any steam being generated, the water and the metal in the pressure vessel must be raised to the boiling point. Since the boiler pressure is at atmospheric pressure, steaming will commence soon after the water reaches 212 F. The heat energy which is added to raise the temperature of the water and metal to the boiling point we know to be sensible heat. At 212 F and atmospheric pressure, the water is at the saturated temperature and is the condition where water and steam can coexist. Furthermore, at the saturation temperature enough heat energy must be added to overcome the latent heat requirement.

Only then will steam start being generated. This period where boiling is taking place and the boiler pressure vessel is starting to build steam pressure is referred to as "**steaming**." Steaming will continue until the boiler reaches set pressure or a pathway for steam to exit the boiler is established. Once steam is allowed to exit the boiler, steam formation is steady until the fuel energy input is stopped. Knowing the boiler mass, water volume, estimated efficiency, and the input fuel rate, one can calculate roughly how long it will take a boiler to make steam from a cold start up.

Example 4.1
A boiler manufacture indicates their boiler weighs 2750 lb empty and contains 125 gallons of water. The fuel input is 1 MMbtu/h and boiler efficiency is 85% If we set the operating pressure at 120 psig, how long will it take for this boiler to make 10 lb/ min steam at the operating pressure if the starting boiler water temperature is 80 F?

*Answer: We can look the steam table in Appendix A and see that the saturation temperature of steam at 120 psig is about 350 F. Furthermore, the heat capacity charts in Appendix A also show water to have a specific heat capacity of about 1 btu/lb-F and carbon steel metal to have a heat capacity of 0.12 btu/lb-F. The boiler pressure vessel metal will therefore require 2750 lb × 0.12 btu/lb-F × (350 − 80)F = **89,100 btu**. The water in the vessel will require 125 gal × 8.3 lb/gal × 1 btu/lb-F × (350−80)F = **280,125 btu**. Altogether 89,100 + 280,125 = 369,225 btu of sensible heat are required to bring the boiler up to the saturation temperature. Since the boiler has a rated input of 1 MMbtu/h and an efficiency of 85%, then the time to transfer 369,225 btu to the water and vessel will be (369,225 btu/1,000,000 btu/h) ÷ 0.85 = **0.43 h or 26 min**.*

*Furthermore, the steam tables indicate that to convert one pound of water at 350 F, 120 psig to steam we need to add 871 btu/lb The boiler output needed is 10 lb/min steam; therefore, the amount of energy required to produce 10 lb/min steam rate at the saturation temperature can be calculated by multiplying 10 lb/min × 871 btu/lb = 8710 btu/ min. Since the boiler input is 1,000,000 btu/h or 16,667 btu/min and at 85% efficiency, 14167 btu/min are useable, we can produce the 10 lb of steam in **less than one minute**. The total time from cold start up to 10 lb/min steam output is simply the time to add enough sensible heat and latent heat or 26 + 1 = **27 min**.*

Note: A boiler on startup takes the bulk of the fuel energy input to heat the water up, and once you reach the saturation temperature, steaming occurs relatively quickly.

The example above implies that the water temperature in the boiler is 80 F when fuel energy is first added. What if the water in our boiler was already at 150 F? Then less sensible heat is required to bring the water up to the saturation temperature. We can calculate the lower amount of sensible heat required to bring the 150 F water and the boiler pressure vessel to the same 350 F saturation temperature as follows:

$$\left(2750 \text{lbs} \times 0.12 \text{ btu/lb-F}\right) + \left(125 \text{gal.} \times 8.3 \text{lbs/gal} \times 1 \text{ btu/lb-F}\right) \times$$
$$\left(350 - 150\text{F}\right) = 273,500 \text{ btu}/0.85 = 321,765 \text{ btu}$$

The difference in sensible heat required between 80 F and 150 F boiler water is 369,225 − 321,765 = 47,460 btu. The 47,460 btu represents a significant savings. Thus, the higher the initial temperature of the water in the boiler, the less heat input needed to bring it to saturation point and, therefore, the less the amount of fuel required to make the water boil. The heat-up time is reduced to 23 min with the 150 F boiler water. This is the principle lesson energy managers convey as they advocate to have all the hot condensate returned to the boiler room instead is discarding it.

Latent and Sensible Heat Versus Pressure

We just calculated the amount of heat energy required to make steam at a specified pressure from water at a specified temperature. The steam generated in the boiler has two portions of heat. These are the sensible heat and the latent heat. Adding the two together, we arrive at the total heat of steam.

$$\textbf{Thus, total heat} = \textbf{Sensible heat} + \textbf{Latent heat}$$

From the steam table in Appendix A we can find each heat value.

Steam Pressure	Temperature	Sensible Heat	Latent Heat	Total Heat
0 psig	212 F	180 btu/lb	970 btu/lb	1151 btu/lb
120 psig	350 F	322 btu/lb	871 btu/lb	1194 btu/lb

The total heat of each lb of saturated steam at 120 psig (vs. 0 psig steam) has increased, but only slightly (by 43 btu). The sensible heat has **increased** by 142 btu, whereas the latent heat has **decreased** by 99 btu.

The basic rules that apply to saturated steam up to about 600 psig for heat versus pressure are

(i) **When steam pressure increases, the total heat increases slightly, the sensible heat increases and the latent heat decreases.**

(ii) **When steam pressure decreases, the total heat decreases slightly, sensible heat decreases and the latent heat increases.**

Note: For steam pressures over 600 psig, the latent and sensible heat relationship follows the same pattern but the total heat decreases.

Thus, **the lower the steam pressure, the lower the steam temperature, but the greater the latent heat.** The significance this concept will become apparent in the next section when we consider the condensation of steam.

CONDENSATION OF STEAM

When steam leaves the boiler, it begins to give up some of its heat to any surface at a lower temperature. In doing so, the steam condenses into water at the same temperature. The process is the exact reverse of the change from water to steam which takes place in the boiler when heat is added. It is the latent heat which is given up by the steam when it condenses. Condensation occurs in the steam piping and in the steam using equipment. The rate of condensation is directly related to the rate of heat absorption of the product being heated or the rate of radiant and convective heat loss in the piping. This is why steam lines should be insulated to prevent massive heat losses and unwanted condensate formation. Let us consider exactly what happens when steam is put to work in a process heating system. Figure 4.2 below shows a jacketed kettle heated vessel which might be found in a typical process steam system.

The vessel is filled with the product, to be heated, and steam is admitted to the space between the outer and inner jacket. The steam then gives up its latent heat to the metal wall of the vessel, which transfers the heat energy to the product. Water is formed as the steam condenses, and runs down to the bottom of the jacketed vessel. This water also called "condensate" must be drained away. When the condensate is drained away, more steam is added until the product is heated to the desired temperature. If the steam in the jacket condenses at a faster rate than the condensate is able to drain away, the bottom of the jacket will begin to fill with water. We call this **water logging**. Water logging may greatly reduce the effectiveness of the steam jacket. While the sensible heat in the condensate is usable energy, maximum heat transfer is obtained if the water is removed from the jacket as quickly as possible, making room for more steam. Steam trap application and sizing plays a very important role in regulating condensate removal.

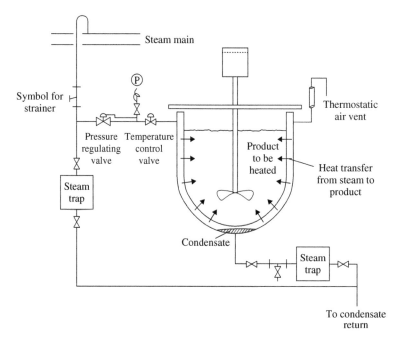

FIGURE 4.2   Typical jacketed kettle steam application.

We can see from the steam charts that the lower the steam pressure the more latent heat is available to be transferred to a product. At 12 psig, 1 lb of steam will surrender 950.1 btus to the product being heated as latent heat, but only allow the temperature to reach a maximum of 243.7 F. Likewise, steam at 120 psig will only surrender 871 btus of latent heat per pound of steam but allow the temperature of the product to rise to almost 350 F. This shows that lower pressure steam will actually transfer more latent heat to the product per pound of steam and is therefore a more efficient energy fluid than high-pressure steam. **Steam system designers should design to use the lowest steam pressure that will allow the product to reach the desired temperature**. This is why steam used to generate hot water is generally low pressure steam.

THE FORMATION OF FLASH STEAM

When hot pressurized condensate or boiler water is released to a lower pressure, its temperature must very quickly drop to the boiling point corresponding to the lower pressure. The surplus sensible heat is utilized as latent heat causing some of it to reevaporate into steam. This reevaporated water is commonly referred to as "flash steam" and is useable even at low pressure. Flash steam charts like the one shown in Table 4.1 below will show the amount of flash steam created when hot pressurized water is subjected to a pressure drop.

The amount of flash steam which each pound of hot pressure water will release can also be calculated. Subtracting the sensible heat of the hot water at the lower pressure from that of the higher pressure water will give the amount of heat available from each pound to provide latent heat of vaporization. Dividing this amount by the actual latent heat per pound at the lower pressure will give the proportion of the hot water which will flash off. Multiplying by the total quantity of hot water being considered gives the weight of low pressure steam available.

Example 4.3
2000 lb/h of condensate at 100 psi is flashed to 10 psi. How much flash steam will be produced. We can see from the steam charts in Appendix A

> *Sensible heat at 100 psi = 309 btu/lb*
> *Sensible heat at 10 psi = 208 btu/lb*
> *Heat available for flashing = 101 btu/lb*
> *Latent heat at 10 psi = 952 btu/lb*
> *Proportion evaporated = 101 divided by 952 = 0.106 or 10.6%*

*Answer: Flash steam available = 0.106 × 2000 lb/h = **212 lb/h***

Another example would be if we had 100 psig condensate being discharged from a steam trap to an atmospheric, gravity flow condensate return system (0 psig), the

TABLE 4.1 Percent Flash Steam

Condensate/Boiler Water Pressure (psig)	Flash Pressure (psig)								
	0	5	10	20	30	40	60	80	100
5	1.6	0							
10	2.9	1.3	0						
15	3.9	2.4	1.1						
20	4.9	3.3	2.1	0					
30	6.5	5	3.7	1.7	0				
40	7.8	6.3	5.1	3	1.4	0			
60	10	8.5	7.3	5.3	3.7	2.3	0		
80	11.8	10.3	9.1	7.1	5.5	4.2	1.9	0	
100	13.3	11.8	10.6	8.7	7.1	5.8	3.5	1.6	0
125	14.9	13.5	12.3	10.4	8.8	7.5	5.3	3.4	1.8
150	16.3	14.9	13.7	11.8	10.3	9	6.8	4.9	3.3
200	18.7	17.3	16.2	14.3	12.8	11.5	9.4	7.6	6
250	20.8	19.4	18.2	16.4	14.9	13.7	11.5	9.8	8.2
300	22.5	21.2	20.2	18.2	16.8	15.5	13.4	11.7	10.2
350	24.1	22.8	21.7	19.9	18.4	17.2	15.1	13.4	11.9
400	25.6	24.2	23.1	21.4	19.9	18.7	16.7	15	13.5

Data from Watson McDaniel [18].

flash percentage of the condensate from Table 4.1 would be 13.3% of the volume discharged.

Conversely, if we had 15 psig saturated steam discharging to the same (0 psig) atmospheric gravity flow return system, the percentage of flash steam would be only 4% by volume. These examples clearly show that the amount of flash released depends upon the difference between the pressures upstream and downstream of the trap and the corresponding temperatures of those pressures in saturated steam. **The higher the initial pressure and the lower the flash recovery pressure, the greater the quantity of flash steam produced**. Flash steam is also generated in boiler when it is rapidly depressurized. This condition is discussed in more detail in the next section.

STEAM ACCUMULATION AND STORAGE

Certain steam using equipment, when initially placed in service, will use steam at a rate far in excess of the boiler's maximum output. The cold metal in the steam using equipment can transfer energy from the steam source as fast as the steam can be supplied to the equipment. Consequently, under these conditions the steam demand on the boiler is very high and the boiler may rapidly depressurize. A boiler that is rapidly depressurized can suffer poor steam quality and nuisance low water shutdowns.

Boiler depressurization rate is a function of the steam demand and the steam system's ability to create instantaneous steam. The amount of instantaneous steam a boiler system will produce is simply the amount of stored steam the boiler will release when the steam header starts to depressurize. Furthermore, the amount of stored steam in the boiler can be expressed as the sum of the steam stored in the steam chest (accumulated dry steam) and the amount of steam stored in the pressurized boiler water (accumulated flash steam).

$$\textit{Boiler instantaneous steam flow} = \textit{accumulated dry steam}$$
$$+ \textit{accumulated flash steam}$$

Accumulated dry steam. An important factor in instantaneous boiler response is steam chest volume. The steam chest volume or steam space is the area in the pressure vessel not occupied by water at the top of the boiler. All pressurized steam piping and boiler steam space will maintain a reservoir of steam. If you determine the volume of this area you can use steam specific volume data to determine the mass of steam stored.

Example 4.3
If a boiler has a steam chest volume of 5 cubic feet and there is 100 ft of 1.5 inch schedule 40 steam piping (0.014 sq. ft cross-sectional area), the total steam chest will be 5 cubic feet + (100 ft × 0.014 square ft.) = 6.41 cubic ft. At 125 psig steam has a specific volume of 3.23 cubic ft./lb. Therefore, the total mass of steam stored is 6.41/3.23 = 2.0 lb of steam.

This amount of stored dry steam in the steam chest is relatively small compared to the stored flash steam of the boiler water. However, the steam chest volume plays an important role in the rate of boiler depressurization and subsequent steam quality.

Accumulated flash steam. A significant amount of stored steam is contained in the pressurized hot water in the boiler. When a boiler is operating at a set pressure, the boiler water is also at that pressure and will release steam, as flash steam, rapidly when it is depressurized. To determine the amount of flash steam one needs to know the upper boiler operating pressure limit and the minimum load pressure limit. The upper boiler operating pressure limit is the pressure which shuts off the boiler burner (i.e., call for heat satisfied). The minimum load pressure is the lowest pressure required to operate the equipment requiring the steam. The difference between the two pressure levels will provide you with the useful instantaneous pressure drop the boiler will experience when a load steam valve opens. This sudden decrease in pressure will cause the boiler water to rapidly boil and produce steam similar to when hot pressurized condensate is flashed in a flash tank.

Using standard flash steam tables like the one shown in Table 4.1 and the water volume (manufactures data) in the boiler, one can easily determine the amount of steam the boiler will surrender when rapidly depressurized.

Example 4.4
*If the same boiler used in Example 2.4 has a water volume of only 77 gallons and maintains the header pressure of 125 psig, is depressurized to 80 psig (minimum useful load pressure), 3.4 percent of the boiler water will flash to steam. The instantaneous steam flow from flash steam therefore is 3.4% × 77 gallons × 8.34 lb/ gallon = **21.8 lb of steam**.*

If the boiler was depressurized to less than 80 psig, even more instantaneous steam would be produced; however, since any steam below 80 psig is not useful to the load equipment, it should not be considered available. Obviously, the water volume of a boiler is critical to responding to sudden load changes. Generally, boilers with high water volumes are more able to handle steam load swings because the larger water volume will produce more flash steam when depressurized.

5

STEAM QUALITY: IT MATTERS

Steam quality is very important but many times unrecognized as a detriment in a steam system. Producing 100% steam from the boiler and maintaining that quality throughout the steam piping is not generally feasible. Water droplets from the boiler and condensate traveling down the steam piping can become entrained. Furthermore, other contaminates like scale, rust, and water treatment chemicals entrained in the water droplets will impact steam quality. This chapter reviews the importance of steam quality and the design and operational practices that influence steam quality. Wet, dry, clean, pure, and superheated steam concepts are defined.

When steam bubbles form in the boiler they float to the water surface, collapse, and release steam. When thousands of these bubbles collapse, microscopic spray droplets are propelled into the steam space above the water. You can observe a similar effect by placing your hand above the surface of a freshly poured glass of soda. In the glass of soda the carbon dioxide bubbles collapse they will spray water droplets on your hand. In a boiler, like the glass of soda, the concentrated spray will form a foam layer at the water surface. The foam layer can vary in size and have a drastic influence on steam quality. A thick foam layer increases the likelihood of water entrainment in the steam and poor steam quality. The entrainment of boiler water in the steam is a condition is often referred to as "**carryover**." Similarly, any steam traveling through the steam piping that comes into contact with a cool area will condense and collect in the bottom of the piping. Turbulent flow in the steam lines can also entrain some of this condensed water. The entrained water in the steam is undesirable and should be removed via steam filters, steam separators, and drip legs.

Process Steam Systems: A Practical Guide for Operators, Maintainers, Designers, and Educators,
Second Edition. Carey Merritt.
© 2023 John Wiley & Sons, Inc. Published 2023 by John Wiley & Sons, Inc.

WHY STEAM QUALITY IS IMPORTANT

Many manufacturing techniques use steam to add energy to a process. Those techniques use equipment designed to have steam quality at or very close to 100%. Since this is very hard to achieve, steps should be taken to ensure the highest steam quality is maintained. Poor steam quality can result in four consequences [3].

1. **Product contamination**. Direct injection steam system for autoclaves, sterilizers, and manufacturing processes can cause contamination of the product with scale, rust, and dissolved solids left behind when the carryover boiler water evaporates. This is especially an issue for steam cooking of food products or sterilization using direct injection. Can you imagine a doctor picking up an operating room utensil that has contamination spots on it left from poor quality steam used in the sterilizer. In these applications, steam filtering is generally required.

2. **Poor heat transfer**. The design of heat transfer equipment is based on high-quality saturated steam. Entrained water reduces the available enthalpy from the steam and can negatively affect performance for heat exchangers, reboilers, kettles, etc. by enhancing water logging. In addition, entrained water can promote water film formation on the heat transfer equipment and cause a loss in efficiency. Later in this chapter, the effect of entrained water on efficiency is quantified.

3. **Damage system equipment**. Entrained water in steam will flash as it passes through a pressure reducing valve (PRV) or control valve. The flashing can erode valve internals and cause premature failures. In addition, steam turbines cannot tolerate entrained water or silica in the steam supply or premature failure can occur. Likewise, the risk of water hammer is higher with poor steam quality. Entrained water can precipitate out in steam lines causing a water slug to form. This water slug can cause water hammer as it changes direction while moving quickly though long pipe runs. The consequences of water hammer are damaged pipes and system components in addition to the noise annoyance. Process steam system can see steam velocities >10,000 ft/min and a slug of water at that velocity can have a lot of kinetic energy.

4. **Overtaxation of boiler feed equipment**. Steam boilers are rated for steam production and any entrained water in the steam effectively increases the boiler's water usage. This can overtax the feed water pumps, steam traps, and water treatment systems. Low-pressure steam systems are especially susceptible to this consequence.

POOR STEAM QUALITY CAUSES AND CURES

There are four major design/operational conditions that affect steam quality: boiler design, steam system pressure management, drip leg efficiency, and water chemistry influence. To produce high-quality steam, all of these conditions must be properly controlled [19].

Boiler design: The first influence on steam quality is the boiler design itself. Steam disengagement area and steam exit velocity will affect steam quality. **Boilers with larger disengagement areas (area of surface water where steam is released) will generally create higher steam quality**. Likewise, the boiler steam nozzle size needs to be large enough to yield exit steam velocities of <5000 ft/min or the foam layer can be pulled right out of the boiler. Because steam-specific volume changes with pressure, manufactures of boilers must ensure the nozzle size is appropriate for the operating pressure. A boiler steam nozzle sized for 120 psig operating pressure will not work well if the boiler is operated much lower than 120 psig. **Engineers should always specify steam exit velocities shall not exceed 5000 ft/min for the range of desired operating pressures**. Steam boilers can be rated for a variety of operating pressures; however, many boilers are rated at either 15 or 150 psig. Care must be taken when a 150 psig boiler is operated at a lower pressure, i.e. 50 psig. Boilers with low steam disengagement areas or high exit velocities will likely require some sort of steam separation downstream of the boiler to yield high steam quality. Table 5.1 shows a chart for boiler nozzle sizes for a variety of pressures that will yield steam exit velocities 5000 ft per minute or less.

Example 5.1
You are operating a 200 hp boiler at 125 psig for a process application. An energy engineer indicates that the boilers will operate more efficiently at 50 psig. What are the consequences of lowering the boiler operating pressure? The boiler steam nozzle is a schedule 40, four (4″) inch nozzle.

Answer: Using specific volume data for 125 psig steam and calculating the inside area of a schedule 40 nozzle, a 200 hp boiler will generate about 7000 lb/h steam flow. At 125 psig the 7000 lb/h or 117 lb/min will yield a 4″ schedule 40 steam nozzle exit velocity of 4500 ft/min and at 50 psig the same steam flow will be about 9000 ft/ min. Lowering the operating pressure to 50 psig will create high exit velocities and poorer steam quality.

Pressure management effects: System pressure influences the foam layer and steam quality. The pressure in the boiler will influence bubble size. Boilers operating at low pressure will create large bubbles and a large foam layer. This is one reason low operating pressure boilers are more susceptible to carryover than high-pressure boilers. Furthermore, when a boiler is rapidly depressurized the steam bubbles rapidly grow in size as they float to the water surface and expand the water volume displacement in the boiler. Consequently, the foam layer (or spray area) rapidly grows in depth, boiler water entrainment occurs, and steam quality suffers. Rapid boiler depressurization can result from high instantaneous load demand caused by quick opening downstream control valves. **Steam systems designers need to ensure appropriate valves and regulators are used to prevent rapid boiler depressurization if high steam quality is important**. This effect is shown below in Figure 5.1. Notice in Figure 5.1 the end result of a rapid boiler depressurization event is not only poor steam quality, but a high potential for a low water boiler shut down condition.

TABLE 5.1 Recommended Steam Nozzle Size (Inches) At Various Operating Pressures

Operating Pressure (psig)	Boiler Horsepower																		
	15	20	30	40	50	60	80	100	125	150	200	250	300	350	400	500	600	700	800
15	4	4	4	6	6	6	6	8	8	8	10	10	12	12	12	12	12	12	12
20	3	3	4	4	4	4	6	6	6	8	8	10	10	10	12	12	12	12	12
40	2	2	2.5	3	3	4	4	6	6	6	6	8	8	8	10	10	10	12	12
60	1.5	2	2	3	3	3	4	4	4	6	6	6	8	8	8	8	10	10	10
80	1.5	1.5	2	3	3	3	3	4	4	4	6	6	6	6	8	8	8	10	10
100	1.5	1.5	2	2	3	3	3	4	4	4	6	6	6	6	6	8	8	8	10
120	1.5	1.5	2	2	3	3	3	4	4	4	4	6	6	6	6	8	8	8	8
135	1.5	1.5	2	2	2.5	2.5	2.5	3	3	4	4	6	6	6	6	6	8	8	8
150	1.5	1.5	2	2	2.5	2.5	2.5	2.5	3	3	4	4	6	6	6	6	6	8	8
200	1.5	1.5	2	2	2.5	2.5	2.5	2.5	2.5	3	4	4	4	4	6	6	6	6	6
250	1.5	1.5	2	2	2	2	2	2	2.5	2.5	3	4	4	4	4	6	6	6	6

Data from Cleaver-Brook Inc. [20].

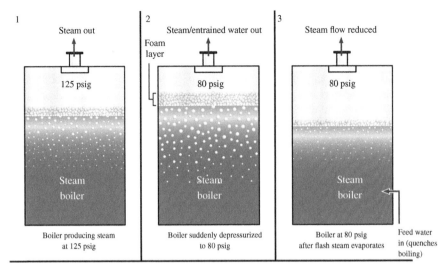

FIGURE 5.1   Effect of rapid boiler depressurization.

Drip leg functionality effects: An often overlooked design criteria is steam piping drip leg design. Poor design can lead to condensate build up and increase the potential for water hammer and water entrainment in steam. Drip legs allow any condensate to drain out of the steam lines. Drip legs need to be installed frequently along the mains, sized appropriately and trapped correctly to be effective. Specific design criteria details are shown in Chapter 9.

Water chemistry influence: Some impurities of boiler water, especially high alkalinity and organics (like oil or soaps) change the water surface tension and will allow small bubbles to collapse into larger bubbles at the water surface. The larger bubbles will create a larger foam layer and consequently more carryover and poorer steam quality. It is important to maintain consistent boiler water chemistry within the manufactures limits.

It should be apparent that a steam boiler will provide the highest quality of steam if the boiler is designed right for the operating pressure and it is operated under steady-state conditions with good water quality.

STEAM CLASSIFICATIONS

The steam tables show the properties of what is usually known as "**dry saturated steam**." This is steam which has been completely evaporated, so that it contains no droplets of water. In practice, boilers will produce steam that often carries tiny droplets of water with it and cannot be described as dry saturated steam. Well-designed and -operated high-pressure boilers will produce steam that is close to dry saturated steam. It is important that the steam used for process heating have little entrained water.

Steam quality is described by its "dryness fraction" – the proportion of completely dry steam present in the steam being considered. The steam becomes "wet" if water droplets in suspension are present in the steam. For example, steam with a dryness fraction of 0.99 means it contains up to 1% moisture. Well-designed and -operated high-pressure boilers can yield dryness fractions of >0.99 and low-pressure boilers can yield dryness fractions of 0.97 or higher. Interestingly, it is the water droplets in suspension which make wet steam visible. Steam is a transparent gas but when subject to the atmosphere, the droplets of water give it a white cloudy appearance due to the fact that they reflect light. **Wet steam** is a term that is used to describe dryness fractions less than 0.97 and is mostly associated with a low pressure or improperly designed or operated steam system. The formula used to determine the dryness fraction is:

Dryness fraction = mass of steam / mass of steam + mass of water in the steam

The enthalpy of wet steam can also be determined once you know the dryness fraction. The steam and saturated water tables can provide the enthalpies of each. Using this data and the formula below, the actual enthalpy of the wet steam can be determined [13].

$$h_t = h_s \times DF + (1 - DF)h_w$$

where h_t = enthalpy of the wet steam (btu/lb)
  h_s = enthalpy of dry steam (btu/lb)
  h_w = enthalpy of saturated water or condensate (btu/lb)
  DF = dryness fraction

Example 5.2
You have determined that the steam in your system has a dryness fraction of 0.96. What is the actual enthalpy compared to dry saturated steam (DF = 1)? The boilers are operating at 50 psig.

*Answer: From the steam tables in Appendix A, we can see that dry saturated steam has an enthalpy of 1179 btu/lb (h_s). From the saturated water tables, we can see that water at 50 psig has an enthalpy of 269 btu/lb (h_w). Therefore, the actual enthalpy of the steam will be Ht = 1179 btu/lb × 0.96 + (1 – 0.96) × 269 btu/lb = **1143 btu/lb**. The wet steam has a 36 btu/lb lower enthalpy than 100% dry saturated steam.*

Clean or pure steam. Often the term clean steam or pure steam is used to describe a level of steam quality. Unfortunately, all types of contaminates like dissolved solids, water treatment chemicals, and corrosion products in addition to water can become entrained in the steam. These contaminate concentrations in steam are usually directly related to the dryness fraction. The more entrained water, the lower the dryness fraction and the more contaminates in the steam. Clean or pure steam is steam that contains very low levels of these types of contaminates. Certain steam applications used in electronics, food preparation, or pharmaceutical applications require clean or pure steam. Food grade direct injection steam is sometimes called 3A quality steam.

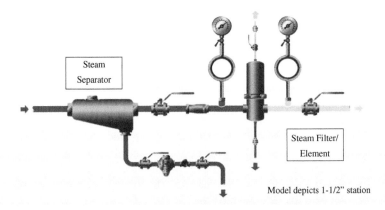

Model depicts 1-1/2" station

FIGURE 5.2   Typical clean steam filter application design [22, 46], Steam Filter Stations for Sterilization, Bulletin SB-P230-01-US-ISS1, www. Spirax Sarco.com.

Two distinct methods exist to achieve clean or pure steam. One method uses ultra-pure water in a stainless steel boiler with a steam separator or filter. This method is limited to unfired steam generators and electric boilers. This technology will provide extremely high level of steam purity but is quite expensive relative to the alternative method of clean steam generation. This method is a good choice if a surplus of "house" high-pressure steam exists and lower-pressure clean steam is required. A typical clean steam generator is shown in Figure 7.11.

The other method used to generate clean steam uses a standard carbon steel boiler with specific design features that allow it to operate with little or no water treatment chemicals [22]. An aggressive blowdown schedule is adopted and boiler water dissolved solids are kept at a low level, thus the opportunity for contaminate carryover is less than typically seen in a conventional steam system application. Steam separators or filters are used to remove entrained water and other contaminates to help enhance steam quality. In addition, complete feed water deaeration is used to minimize the use of oxygen scavenger-type water treatment chemicals. This method of clean steam generation is more cost effective than the previous mentioned method, but may not yield as high a steam quality as with a stainless steel unfired steam generator. Regardless of which method is used to generate the steam, pure steam systems will require steam filtration similar to what is shown in Figure 5.2 above. **For the design engineer, understanding the level of steam quality actually required for a particular application is very important**.

MEASURING STEAM QUALITY

Measuring steam quality accurately is not easy. There are sophisticated measuring systems that do measure steam purity; however, their practical use is limited to a laboratory setting. For most steam system operators that wish to determine the general steam quality their boiler is generating, there is a method that is relatively simple and easy to implement. The principle of this method is to determine the ratio of impurities in the steam relative to the level in the boiler. Ideally, dry steam should be steam vapor only and any impurities in the steam can be associated with boiler water carryover or contamination from the steam piping. Consequently, steam purity can be measured by condensing the steam in a controlled manner and measuring a dissolved ion found in the boiler water. Conductivity or chloride ion concentrations are often used as the common marker ion. A ratio of steam conductivity or chloride concentration to boiler water conductivity or chloride concentration is used as a crude but adequate indicator of steam purity. See Figure 5.3. Since boiler water chemistry changes, the two measurements needs to be made at steady state and with a minutes of each other. An example is shown below.

Example 5.3
You wish to determine the quality of steam in your system. Using the method described previously you measure the boiler water and find the boiler has a conductivity of 2500 uS/cm², and the condensed steam has a conductivity of 50 uS/cm². What is the steam purity?

*Answer: The steam purity is the ratio of steam conductivity to the boiler water conductivity. Therefore, steam purity = 1 − (50 uS/cm² ÷ 2500 uS/cm²) = **0.98 or 98%**.*

SUPERHEATED STEAM

If heat energy is added to dry steam, the steam temperature will rise. The steam is then called "superheated," and this "superheated steam" will be at a temperature above that of saturated steam at the corresponding pressure. The heat added to dry steam to raise the temperature is sensible heat. To obtain superheated steam, all the entrained water in the steam must first be evaporated. It is possible to use a steam superheater to produce steam several hundred degrees higher than the saturation temperature.

Superheaters can be electric or gas fired or, integral to the boiler design. They provide process engineers a means to provide a high energy steam to a process without the corresponding high pressures required with saturated steam. Some steam boiler types, like the water tube type, can produce superheated steam by recycling the saturated steam back through a section of the furnace. Superheated steam tables are provided for reference in Appendix A.

Some steam applications may require superheated steam. One application where superheated steam may be required is the use of a steam turbine. The turbine's blades require very dry steam because the moisture and associated impurities can destroy the blades. A steam turbine converts the steam energy into rotational energy. The higher energy dry steam allows the turbine to extract more energy.

The condensation of superheated steam can require special precautions [23]. Saturated steam will condense very readily on any surface which is at a lower temperature, so that it gives up the latent heat. However, when superheated steam gives up some of its enthalpy, it does so by virtue of a fall in temperature. No condensation will occur until the saturation temperature has been reached. Because thermal conductance is lower in gases than liquids, it is found that the rate at which we can get energy to flow from superheated steam is often less than we can achieve with saturated steam, even though the superheated steam is at a higher temperature. Superheated steam, because of its other properties, is the natural first choice for power steam requirements, while saturated steam is routinely used for process and heating applications.

Superheated steam can also be created downstream of a pressure reducing valve (PRV). A PRV reduces the steam pressure; however, the steam retains the enthalpy of the higher-pressure steam. An example is shown below.

Example 5.4

Steam at 125 psig passes through a PRV where its pressure is reduced to 60 psig. What is the steam energy content profile of the lower pressure steam?

Answer: From the steam tables in Appendix A, saturated steam at 125 psig and 60 psig and superheated 60 psig (SH) is shown below.

Steam Pressure	Temperature	Sensible Heat	Latent Heat	Enthalpy
125 psig (saturated)	353 F	325 btu/lb	868 btu/lb	1193 btu/lb
60 psig (saturated)	307 F	277 btu/lb	906 btu/lb	1183 btu/lb
60 psig (superheated)	353 F	287 btu/lb	906 btu/lb	1193 btu/lb

The 60 psig superheated steam has a total enthalpy of 1193 btu/lb or 10 btu/lb more energy than dry saturated 60 psig steam. Interestingly, the data also shows that steam has a heat capacity of 10 btu/lb/ 353-307 F = 0.22 btu/lb-F at this temperature range. This is much lower than the 1 btu/lb-F value for water. Consequently, the superheated steam at 60 psig would need to be cooled to <307 F before it will condense and surrender its latent heat.

Steam trapping of equipment that uses superheated steam can be tricky. Two different scenarios must be considered when using superheated steam; the startup load and the steady-state load. Startup conditions will result in some amount of condensate that will need to be removed. Steady-state operations, however, may need to deal with trapping steam vapor. Certain steam traps need to be used for superheated steam applications to address both scenarios. More on the specifics in Chapter 9.

Desuperheaters are sometimes installed if the system was not designed to handle the higher temperature associated with the superheated steam. A desuperheater generally uses condensate or fresh makeup water sprayed into the superheated steam at a mass flow high enough to absorb the sensible heat above the saturation temperature. A typical desuperheater design is shown in Figure 5.4.

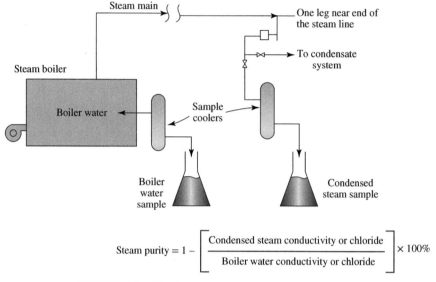

$$\text{Steam purity} = 1 - \left[\frac{\text{Condensed steam conductivity or chloride}}{\text{Boiler water conductivity or chloride}} \right] \times 100\%$$

FIGURE 5.3   Simple steam purity measurement system.

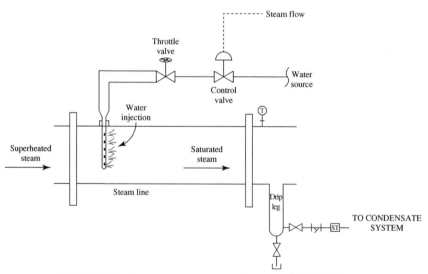

FIGURE 5.4   Desuperheater concept. Courtesy of C&S Engineers.

Example 5.5

In the example above we determined the 60 psig superheated steam would need to be cooled to 307 F to desuperheat it. If the mass flow of this 60 psig steam at 353 F is 1000 lb/h, how much 180 F condensate would need to be added to convert the superheated steam to saturated steam.

*Answer: We know from the steam tables that 60 psig steam at 353 F has an enthalpy of 1193 btu/lb and the same steam at 307 F has an enthalpy of 1183 btu/lb. Therefore, the desuperheat needs to absorb 10 btu/lb of steam. At 1000 lb/h steam flow, the amount of energy available to injected condensate is 10 btu/lb × 1000 lb/h = 10,000 btu/h. When you desuperheat, the goal is to convert all injected water to saturated steam. Likewise, from the steam tables that converting 1 lb of 180 F water to 60 psig steam we will need (307 F-180 F) × 1 btu/lb-F + (905 btu/lb) = 1032 btu. Therefore the mass of 180 F water needed to desuperheat is 10,000 btu/h/1032 btu = **9.7 lb/h** or 1.1 gph.*

6

THE STEAM SYSTEM DESIGN

STEAM SYSTEM TYPES

Generally steam systems can be classified into one of three categories.

1. Steam heating systems
2. Power generation steam systems
3. Process steam systems

Steam heating systems are generally closed loop low-pressure systems found in government and school buildings. The degree of control and efficiency is usually less than what is found in process or power generation applications. Similar comparison can be made with respect to steam quality. Steam heating systems tend to be large and less sophisticated than process or power generation steam systems. Many of these low-pressure steam heating systems as shown in Figure 6.1, are being replaced with higher efficiency hot water heating systems.

One useful steam system found in some low-pressure applications is called the Hartford Loop design. This design is shown in Figure 6.2 below. Notice the absence of any condensate pumps. This design uses the head pressure of the condensate line to push condensate back into the boiler. This was for many years a very simple way to use steam to heat a building.

Power generation steam systems can vary in size. Some are large high-pressure steam systems that use steam to drive a turbine generator as seen in Figure 6.3. Nuclear power, coal, biomass-fired power plants, and other combined cycle cogeneration systems fall into this category. This type of steam system will generally require clean dry or superheated steam for extended turbine life and high purity condensate. Controls can be quite sophisticated (nuclear reactor) or can be quite simple (newer cogeneration steam). Cogeneration may use Heat Recovery Steam Generators or

Process Steam Systems: A Practical Guide for Operators, Maintainers, Designers, and Educators,
Second Edition. Carey Merritt.
© 2023 John Wiley & Sons, Inc. Published 2023 by John Wiley & Sons, Inc.

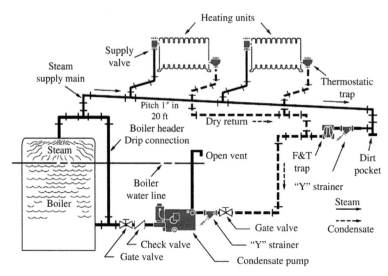

FIGURE 6.1   Steam heating system with pumped returns. Courtesy of Xylem Inc.

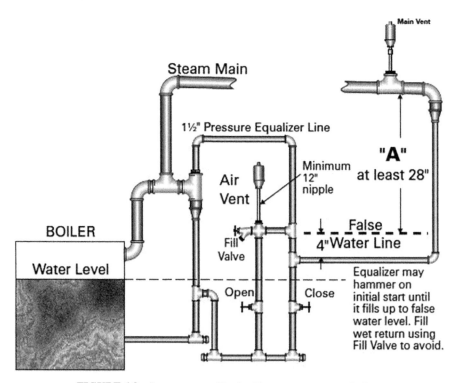

FIGURE 6.2   Low-pressure Hartford loop steam system design.

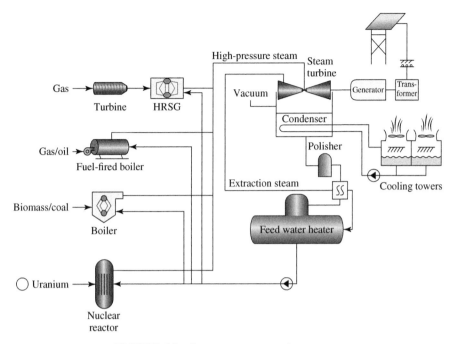

FIGURE 6.3   Steam power generation system.

HRSGs to capture waste heat and convert the energy to usable steam [24]. Steam quality and high system efficiency requirements influence the design of these steam systems. Combined heat and power (CHP) systems are covered in more detail in Chapter 14 in the book. Although this book focuses mainly on process steam system, many of the principles can be applied to steam power generation system.

   Process steam is the type of steam system that is the most applicable for the guidance in this book. A simple schematic of a process steam system is shown in Figure 6.4. These systems can range in size from small (<100 lb/h) to very large (>100,000 lb/h). Steam quantity and quality requirements influence the system design. Process steam systems tend to have very automated controls with tight tolerances. System reliability is essential, and preventative maintenance on these steam systems is critical. The steam system in a process application is usually considered an ancillary system. Steam is required for the process but is not the salable product. Consequently, steam systems may not receive attention until there is a problem. A good example of this application is brewing. Without steam flow, the output of the brewery comes to a halt. A good process steam system operation requires constant visibility with the plant management team.

THE PROCESS STEAM SYSTEM: AN OVERVIEW

The steam generated in the boiler must be conveyed through pipe work to equipment where its heat energy is used in some process-related application. Once the steam condenses and surrenders its energy, the condensate must be removed, collected, and

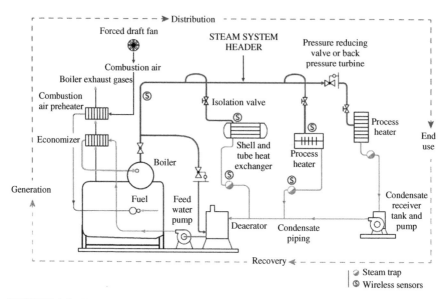

FIGURE 6.4   Process steam system. Courtesy of the Advanced Manufacturing Office, US Department of Energy [25].

prepared for reuse in the boiler. This cycle of steam generation, steam delivery, condensing, collecting, and pumping back to the boiler is the basic concept of a process steam system.

Generally, we can consider the basic steam system to be comprised of five subsystems; the fuel delivery and combustion system, the steam generator, the steam delivery system, and the condensate/feed water system are the four subsystems that account for the production of steam. The remaining subsystem, the water treatment subsystem, does not play a direct role in steam production; however, it plays an extremely important role in maintaining good steam quality and long life of the entire steam system equipment. The five subsystems are shown in Figure 6.5.

The fuel handling and combustion system is the part of the system that controls input to the steam generator. The fuel handling system may be a set of gas valves and regulators or oil valves and regulators for gaseous and liquid fuels. It may also be an auger and conveyor system for solid fuels like wood chips, coal, or refuse. For electric boilers the fuel handling system would be the power feed or element contactors to the steam generator. In all cases, this part of the system controls the input energy and consequently controls the steam output.

The combustion system takes that fuel input and converts it to thermal energy. The efficiency of the combustion system plays a major role in the overall efficiency of the steam generator. This system mixes the right amount of air with the fuel to ensure it combusts completely. Modern combustion systems have feedback controls that tie the steam generator flue gas chemistry to the generator combustion system to ensure high efficiency. In the case of electric boilers the combustion system is essentially the

① Fuel delivery and combustion system

② Steam delivery

③ Condensate collection and feed water

④ Steam generator

⑤ Chemical/water treatment

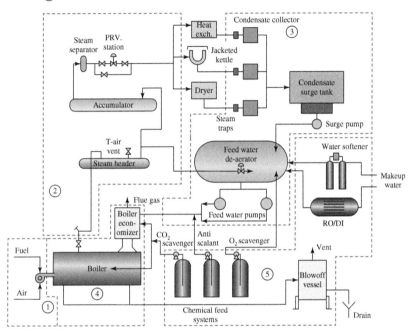

FIGURE 6.5 The five subsystems of a process steam system.

electric elements or electrodes. The efficiency of the electric elements or electrodes is related to their cleanliness. The fuel handling and combustion controls system details are presented in Chapter 8.

The steam generator or boiler is the heart of the steam system. This is the machine that transfers the fuel energy from the combustion system to water in a pressure vessel to make steam. Steam generators come in a variety of sizes and configurations. This part of the system includes the insulated pressure vessel, pressure and water level controls, steam flow, and the combustion flue gas removal. A detailed look at the available types and uses of the many steam generators is covered in detail in Chapter 7.

The steam delivery system is the part of the steam system that conditions the steam to the right quality and pressure, and conveys it to the steam using equipment. This part of the system includes steam piping, pressure reducing and temperature control valves, steam separator and filters, steam accumulators, and steam flow meters. Each steam delivery system is uniquely designed for each process application. More detail is offered in Chapter 9.

The condensate and feed water system is the part of the steam system that separates the steam from the condensed water, collects it, and transfers it back to the steam generator. This part of the system includes the steam traps, condensate piping, condensate collection tanks, feed water tank or deaerator, pumps, and controls. The function of this part of the system is to provide a means to collect and feed hot, high-quality water to the steam generator for reuse to make more steam. Chapters 10 and 11 provide a comprehensive review of the design and operational requirements of this part of the steam system.

The last part of the steam system is the **water treatment system**. This part of the steam system is not directly involved with the steam cycle but is considered essential to maintain reliability and durability of the system components. The water treatment system includes the water conditioning equipment and controls needed to render the makeup water, feed water, and boiler water harmless relative to corrosiveness, scaling, and sludge buildup potential. The steam system contains metal components in contact with water and without proper water conditioning, these metal components would not maintain system integrity very long. Specific methods used to treat steam system water and the consequences of improper treatment are described in Chapter 12.

7

THE STEAM GENERATOR

One of the most important decisions made in any process steam system is the steam generator selection. There are many designs to choose from, and matching the right steam generator type for an application is critical. This chapter reviews the major steam generator designs and provides guidance to help you determine when each type should be the basis for design. In this chapter the term "steam generator" is used interchangeable with the definition of a "steam boiler." Steam generators are designed for low-pressure or high-pressure applications. Low-pressure boilers generally are considered to be a boiler <50 psig design. High-pressure boilers are typically used for process loads and can have an operating pressure of up to several hundred psig. Most process steam systems use saturated steam. Steam loads are usually specified in lb of steam per hour and boilers are marketed with a horsepower rating with 1 hp equal to 34.5 lb/h steam generation. Boilers are available in a variety of sizes, functionality, and physical layout.

Steam boilers are defined according to design pressure and operating pressure. Design pressure, also called the **MAWP or maximum allowable working pressure**, is used for the purpose of calculating the minimum permissible thickness or physical characteristics of the boiler pressure vessel construction. The safety valves are required to be set at or below the MAWP. **Operating pressure** is the pressure of the boiler at which it normally operates. The operating pressure usually is maintained at a suitable level (i.e., at least 10%) below the setting of the pressure relieving valve(s) to prevent their frequent opening during normal operation. The boiler pressure sensors are adjustable and will determine the operating pressure range of the boiler.

Process Steam Systems: A Practical Guide for Operators, Maintainers, Designers, and Educators,
Second Edition. Carey Merritt.
© 2023 John Wiley & Sons, Inc. Published 2023 by John Wiley & Sons, Inc.

THE IDEAL STEAM GENERATOR

Steam generators (boilers) will have both good and not so good design features. In fact, there is no perfect boiler made today. Below is a list of desirable features and why those features are desirable. It is essential that engineers look at these features and choose the best boiler type for each application they encounter.

1. **Small footprint:** Boiler room real estate can be expensive to build, and retrofitting a boiler room with a fixed amount of space will require a boiler that occupies little area.
2. **High water volume:** A boiler with high water volume will have a high stored energy level and be able to respond more effectively to a sudden steam load demand.
3. **Large steam disengagement area:** The larger this area is, the less chance for water entrainment in the steam and the higher the steam quality.
4. **Quick startup:** When you want steam …you don't like to wait.
5. **High mass pressure vessel:** A durable design will be forgiving relative to operator abuse, that is, water treatment, and generally will provide high return on investment.
6. **Good performance over a wide range of operating pressures:** Process applications can require steam pressure from 20 to >400 psig operation.
7. **Simple accessible boiler internal components:** The maintenance folks love this one.
8. **Low heat flux on the waterside and high surface area on the furnace side:** Low heat flux will reduce the waterside scaling potential and the high surface area will aid in higher heat transfer efficiencies.
9. **High burner turndown:** This will reduce cycling, prolong boiler life, and allow the boiler to accommodate low load periods.
10. **Ability to interface with the building management controls:** Allows for remote monitoring and reduced operator interface time.

Some of these features are contradictory. This is why there exists no perfect boiler. The specifying engineer must determine what features are the most important based on the application and select a boiler design that best meets the need. For instance, if the boiler room has little room, then footprint becomes a very important feature. Likewise, if the application requires steam at 600 psig, then only certain boiler types can be used. Problems with a boiler installation can be a consequence of misjudging the boiler design features and how those features affect the overall performance of the steam system. Low-pressure boilers will need high disengagement areas to have any hope of good steam quality. Fortunately, many misapplied boiler installations can be compensated by adapting the steam delivery system design to compensate for the misapplication. Table 7.1 shows some general application ratings for the various types of boilers versus the 10 desirable features.

TABLE 7.1 Boiler-Type Comparisons

	Scotch Marine	Firebox	Watertube	Flex Tube	Cast Iron	Vertical Tubeless	Biomass	Electric Resistance	Electrode	Unfired Steam Generator
Small footprint	P	M	M	M	G	G	P	G	G	M
High water volume	G	M	P	P	P	P	G	M	M	G
Large steam disengagement area	G	M	M	M	P	P	G	M	G	G
Quick startup	P	G	G	G	G	G	P	P	G	M
High mass pressure vessel	G	M	G	P	M	M	G	M	G	G
Good performance over a wide range of operating pressures	M	P	G	P	P	M	G	G	G	G
Simple accessible boiler internals	M	G	P	G	G	M	M	G	P	P
Low heat flux/high surface area	G	M	G	G	M	P	M	P	P	G
Efficient at high turndown	P	P	M	M	M	M	P	G	P	G
BMS interfaces capability	G	G	G	G	G	G	G	G	G	G

G, good; P, poor; M, moderate.

Steam Generator Types

Fundamentally steam generators can be grouped in one of four categories: fuel-fired, electric, solid fuel, or unfired steam generators. The following is an overview of different types of boilers found in each category.

FOSSIL FUEL-FIRED BOILERS

Scotch Marine: The Classic Horizontal Firetube Boiler. The Scotch Marine style of boiler has become so popular in the last 50 years that it frequently is referred to simply as "a firetube boiler." Firetube boilers are available for steam pressures up to 300 psig and are typically used for applications ranging from 60 to 800 hp; however, sizes up to 2000 hp are occasionally manufactured. A scotch marine boiler is a cylindrical vessel, with the flame in the furnace and the combustion gases inside the tubes. The furnace and tubes are within a larger vessel, which contains the water and steam. A cutaway view of a typical scotch marine boiler is shown in Figure 7.1.

The horizontal firetube construction provides some characteristics that differentiate it from other boiler types. Because of its vessel size, the firetube contains a large amount of water allowing it to respond to load changes with minimum variation in steam pressure. However, steam pressure in a firetube boiler is generally limited to approximately 300 psig. To achieve higher pressure, it would be necessary to use an impractical thick shell and tube sheet material. Scotch marine-type boilers are limited to minimum sizes around 60 hp or 2000 lb/h steam output. This type of boiler is well suited for

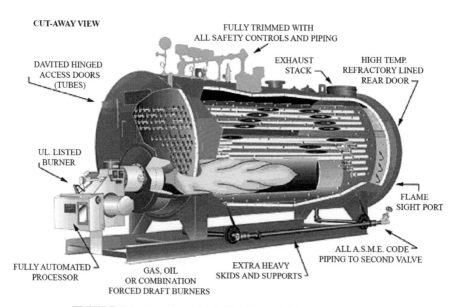

FIGURE 7.1   Scotch marine boiler. Hurst Boiler and Welding Co.

low- to medium-pressure heating and process applications; however, due the pressure vessel orientation and tube removal, it will require a relatively large footprint.

Firetube boilers are usually built similar to a shell and tube heat exchanger. A larger quantity of tubes results in more heating surface per boiler horsepower, which greatly improves heat transfer and efficiency. The furnace pipe (sometimes called the morrison tube) and the downstream banks of tubes carrying the hot furnace gases transfer heat to the water. Combustion occurs within the front section of the furnace, and the flue gases are routed through the tubes to the stack outlet. Scotch marine boilers are available in two-, three-, and four-pass designs. A single "pass" is defined as the area where combustion gases travel the length of the boiler. Generally, boiler efficiencies increase with the number of passes; however, total surface area and turbulent flue gas flow will ultimately determine heat transfer efficiency. The use of turbulators and ribbed type-tubes helps create flue gas turbulence in the boiler tubes. One kind of enhanced boiler tube design are XID-type tubes as shown in Figure 7.2.

Horizontal firetube-type boilers are available in either dryback or wetback design. The construction schematic of both are shown in Figure 7.3 In the dryback design, a refractory-lined chamber, outside of the pressure vessel, is used to direct the combustion gases from one pass to the subsequent pass …, that is, the furnace pipe to the first tube bank. Easy access to all internal areas of the boiler including tubes, burner, furnace, and refractory is available from either end of the boiler. This makes maintenance easier and reduces associated cleaning costs. The wetback boiler design has a water-cooled turnaround chamber used to direct the flue gases from the furnace to the tube banks. The wetback design requires less refractory maintenance; however, internal pressure vessel maintenance, such as cleaning, is more difficult and costly. The wetback design will create less radiant losses than a dryback design and generally have a slightly higher efficiency.

Firebox boilers. The firebox boiler uses similar tube arrangements as the horizontal firetube boiler. Its combustion chamber is a large "boxlike area," unlike the firetube's cylindrical furnace. The firebox boiler, which has a large steam disengagement area, is typically suited for low-pressure steam applications. The firebox boiler is a compact, low capital cost unit and serves as a good fit for seasonal use and when efficiency is not the driving factor. Sizes range is limited to about 600 hp. The use of firebox boilers in the United States is diminishing mainly due to low efficiency and the overall conversion of the low-pressure steam market to hot water systems. The firebox boiler is a good choice if a combination low-pressure steam/hot water ASME section IV boiler is desired.

Industrial watertube boiler. The industrial watertube boiler typically produces steam primarily for industrial process applications. The watertube design is one in which the tubes contain steam and/or water and the products of combustion pass around the tubes. Typically, watertube designs consist of multiple drums as shown in Figure 7.4. A steam drum (upper) and mud or feed water drum (lower) are connected by the tubes, which form both the convection section and the furnace area.

Packaged industrial watertube boilers are typically rated in pounds of steam per hour output at operating conditions and range from 10,000 to >100,000 lb/h.

FIGURE 7.2   XID tube design. Rosa Operating.

Industrial watertube boilers are noted for their fast steaming capability. Steam is generated relatively quickly because of the low water content. The low water content,
however, make water treatment a little more difficult to control and does not provide
as good an instantaneous supply of stored steam as the firetube design. The industrial
watertube boiler design makes it capable of generating either saturated or superheated steam. When applications dictate superheated steam usage, or high pressures
(greater than 300 psig), an industrial watertube boiler should be considered. Watertube
boilers have a smaller footprint than the HFT but can require 20 foot ceilings for the

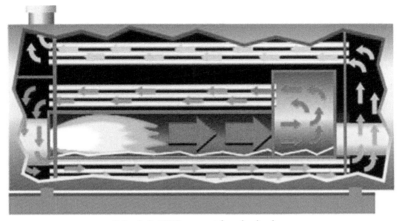

Model 4WI – wet back design

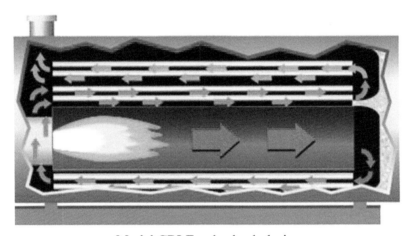

Model CBLE – dry back design

FIGURE 7.3 Dryback versus wetback boiler configuration. Courtesy of Cleaver-Brooks, Inc.

larger units. Repairs on a watertube involve opening up the outer insulation and repairing or plugging a leaking tube. Mud drum design lends itself well to periodical purging of sludge.

Flexible watertube (bent-tube) boilers. Flexible watertube or "flex-tube" boilers are a common type of boiler used for heating applications because of their resistance to thermal shock and high heating surface area. Flexible watertube boilers are available in size ranges up to 300 hp and are available for low- and medium-pressure steam applications. Field erectable packages are also available. These types of boilers have a relatively low capital cost to manufacture and have low water volumes which facilitate quick startups but poor response to changing load demand. Flex-tube boilers

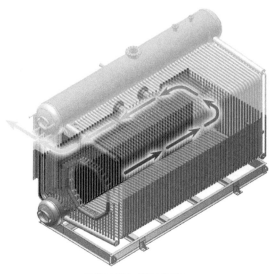

D-Type Water Tube Design

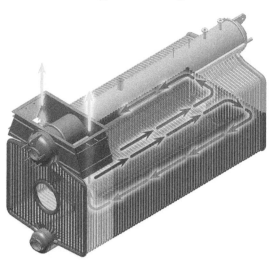

O-Type Water Tube Design

FIGURE 7.4   Watertube boiler design. Victory Energy Operations, LLC.

are primarily heating application-type boilers. A typical Flexible water tube boiler is shown in Figure 7.5.

Cast iron boilers. Cast iron sectional boilers have the boiler water contained in cast iron sections that are bolted together. The furnace gases flow through the middle and over the iron sections. These boilers are popular with schools and older government buildings that used low-pressure steam heating. They have very limited application in the process industry due to the low-pressure limitation. Cast iron

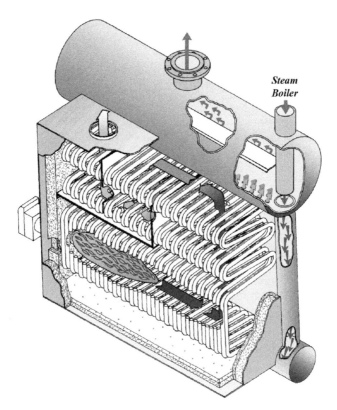

Steam
Boiler

FIGURE 7.5   Flexible watertube boiler. Bryan Boiler Co.

sectional boilers typically range in size from 25 to 200 hp. One advantage of the cast
iron boiler is its modular design, which includes sections for field erection/repair.
Cast iron boiler durability is compromised with thermal shock. These types of boilers
are simple and cost effective when efficiency is not a big factor at initial investment.
Their popularity is diminishing in the United States and mirrors the decline in steam
heating system use in our country. A typical cast iron boiler is shown in Figure 7.6

Vertical tubeless boilers. Vertical tubeless boilers are a variation of a firetube
boiler. They are a single- or multiple-pass boiler that use a pipe or series of pipes within
a pipe as a pressure vessel design as shown in Figure 7.7. The combustion gases flow
down from the top or up from the bottom through an interior furnace pipe. In a one-pass
system the flue gases are exhausted; however, in a two-pass design, the flue gases are
turned at one end of the boiler and return back to the opposite end of the boiler and
exhausted. This type of boiler is limited to about 250 hp in size due to the height that is
required to obtain any level of steam generation efficiency. The major advantages of
this type of boiler design in the small footprint and simple pressure vessel design.
Vertical tubeless boilers are known for their durability and simplicity, making them a
favorite design for the laundry dry cleaning market. One disadvantage of this boiler
design is the smaller steam disengagement area which can lead to steam quality issues

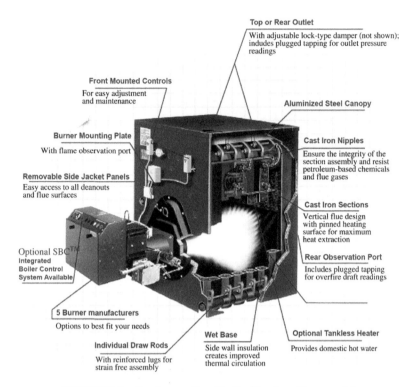

Top or Rear Outlet
With adjustable lock-type damper (not shown); indudes plugged tapping for outlet pressure readings

Front Mounted Controls
For easy adjustment and maintenance

Aluminized Steel Canopy

Burner Mounting Plate
With flame observation port

Cast Iron Nipples
Ensure the integrity of the section assembly and resist petroleum-based chemicals and flue gases

Removable Side Jacket Panels
Easy access to all deanouts and flue surfaces

Cast Iron Sections
Vertical flue design with pinned heating surface for maximum heat extraction

Optional SBC™
Integrated
Boiler Control
System Available

Rear Observation Port
Includes plugged tapping for overfire draft readings

5 Burner manufacturers
Options to best fit your needs

Wet Base
Side wall insulation creates improved thermal circulation

Optional Tankless Heater
Provides domestic hot water

Individual Draw Rods
With reinforced lugs for strain free assembly

FIGURE 7.6   Cast iron sectional boiler. Burnham Commercial.

if boiler water chemistry or steam load demand is not regulated. Vertical boilers make excellent small medium pressure process steam boilers. Boiler manufactures use turbulators and Alu-fer technology (discussed in more detail in Chapter 16) to increase surface area and efficiency. Vertical boilers tend to have low water content which allow for a quick startup but poorer response to load swings.

SOLID FUEL-FIRED BOILERS

Biomass boilers are a variation of the industrial water–tube- or firebox-type boiler. These boilers use wood chips, municipal solid waste, saw dust, coal, or other combustible-type solid material as the fuel. The combustion chamber and fuel feed systems are much different than on a standard gas- or oil-fired water-tube boiler. Solid fuel-type boilers rely on an auger system, a hydraulic ram system, or a manual feed system to push solid fuel into the furnace. The output of the boiler is controlled by the rate that the solid fuel is pushed into the furnace. With the increased interest in biomass as a fuel, a lot of recent research and development has been performed on this type of boiler. A typical biomass boiler is shown in Figure 7.8. Biomass contains varying amounts of moisture and btu content. Consequently, optimization of combustion to yield low emissions and consistent output has been the challenge for

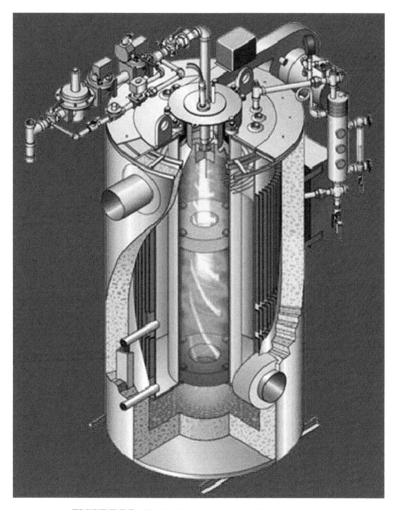

FIGURE 7.7 Vertical tubeless boiler. The Fulton Co.

the boiler manufactures. Biomass boilers tend to be quite large capacity (>5000 lb/h) but have the advantage of using a low-cost industrial waste product. Industries that have available wood by-products, like sawmills and furniture makers, find steam generation via this method very beneficial.

The two drawbacks of this boiler type are the poor ability to handle rapid load changes and the difficulty performing a quick shutdown if the steam load is lost or a boiler safety activates. Unlike a gas or oil fuel valve, the solid fuel in the furnace cannot be easily extinguished or rapidly ramped up. Sometimes steam quenching is used to help put out combustion of the solid fuel. One very good steam generation system will use the biomass boiler as a baseline boiler and a fuel-fired boiler as a topping boiler. The fuel-fired boiler can handle the load swings. The biomass boiler is sized to handle the minimum steam load.

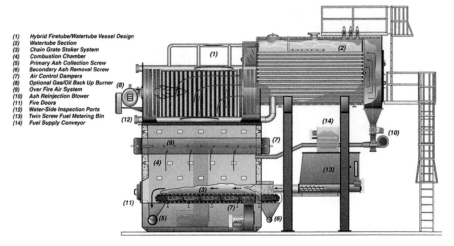

(1) Hybrid Firetube/Watertube Vessel Design
(2) Watertube Section
(3) Chain Grate Stoker System
(4) Combustion Chamber
(5) Primary Ash Collection Screw
(6) Secondary Ash Removal Screw
(7) Air Control Dampers
(8) Optional Gas/Oil Back Up Burner
(9) Over Fire Air System
(10) Ash Reinjection Blower
(11) Fire Doors
(12) Water-Side Inspection Ports
(13) Twin Screw Fuel Metering Bin
(14) Fuel Supply Conveyor

FIGURE 7.8   Biomass boiler by the Hurst Boiler and Welding Co.

ELECTRIC BOILERS

Electric boilers can be either resistance or electrode type. Both are noted for being clean, quiet, easy to install, and compact. Because there are no combustion considerations, an electric boiler has minimal complexity (no fuel or air handling equipment). They are built with replaceable heating elements. An electric boiler may be the perfect alternative to supply steam where the customer is restricted by emission regulations or access to gaseous and liquid fuel is limited. In areas where the cost of electric power is minimal, the electric boiler might also be the best choice. One important fact about electric boilers is their ability to be fabricated with stainless steel pressure vessels. Unlike fuel-fired boilers, ASME code allows electric boiler to be constructed of the higher-grade metal which when combined with high-quality water will produce very high-quality steam. Small point of use boilers are a good fit for hospitals, pharmaceuticals, electronic component manufactures, and R&D labs.

Resistance-type electric boilers use immersion elements to heat up the boiler water to make steam. The metal elements have a high electrical resistance and will heat up fast when current is passed through them. This works similar to the hot wire found in your toaster. The elements must be surrounded by water or they will burn out quickly. By controlling the amount of current supplied to each element, the output of the boiler can be regulated. Sizes range from 1.5 to 200 hp output. Resistance element boilers actually respond more quickly than fuel-fired boilers. Because there is no combustion of fuel, the temperature of the vessel never exceeds the temperature of the water or steam. The combustion section of a fuel-fired boiler can see temperatures above 2000 °F, which is why they must warm up more slowly than electric boilers. Some areas of the country restrict the use of commercial electric boilers.

Electric boiler elements have a very high heat flux and will be susceptible to scaling if water chemistry is not maintained. Poor water chemistry can also cause scaling and poor heat transfer efficiency in fuel-fired boilers. Figure 7.9 is a picture of a typical electrical steam boiler.

Electrode-type electric boilers provide a unique method of steam generation. Basically an electrode boiler is a vertical pressure vessel containing a set of metal electrodes suspended from the top head that have a very high voltage applied to them. Steam is generated by pumping water to the top of the vessel where it flows through nozzles that impact the electrodes with the water stream serving as resistive conductors. The current flowing through the water vaporizes a portion of the water to steam. The amount of steam generated is proportional to the flow rate of boiler water striking the electrodes, the level of conductivity in the boiler water, and the voltage applied to the electrodes. A typical electrode boiler is shown in Figure 7.10

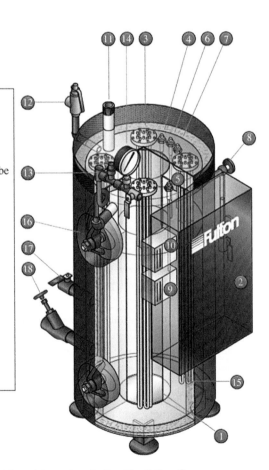

1. Pressure vessel is built to
   ASME code
2. Electrical control panel box
3. Electric heating elements
4. Low water cutoff probe
5. Auxiliary low water cutoff probe
6. Pump "On" probe
7. Pump "Off" probe
8. Sight glass assembly
9. Operating pressure control
10. High limit pressure control
11. Steam outlet
12. Safety valve
13. Steam gauge assembly
14. Steam pressure gauge
15. High-temperature insulation
    surrounds the vessel
16. Large (3″ × 4″) easily
    access handholes
17. Feedwater shutoff valve
18. Blowdown valve

FIGURE 7.9   Electric resistance-type boiler. The Fulton Co.

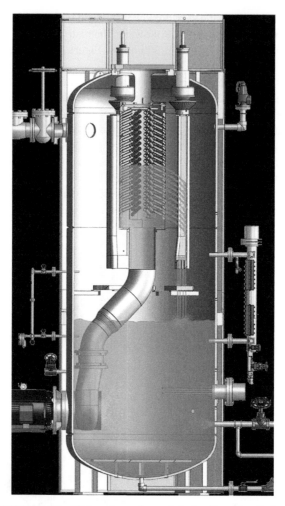

FIGURE 7.10   High-voltage electrode boiler. Precision Boilers.

These types of boilers ranging up to 3000 hp use a lot of electricity and are a favorite auxiliary boiler for power generation plants.

Electrode boilers produce steam faster than any other boiler design. Electrode boilers offer infinite modulation, adjusting steam output quickly and precisely to the process demand. One drawback of these types of boilers is chemistry maintenance required on the boiler feed water. Boiler feed water is typically treated to maintain proper pH, conductivity, and to "scavenge" oxygen that leads to corrosion inside the vessel and piping system. Electrode boilers require more attention to water treatment than other boiler types. These operate with increase boiler water conductivity which is required to increase the boiler water conductance and allow for the vaporization of water to steam. Likewise, water impurities must be kept within limits to prevent those impurities from precipitating inside the boiler and/or coating the electrodes. Iron corrosion products must be removed from the boiler.

UNFIRED STEAM GENERATORS

Unfired steam generators are boilers that use an existing heat source to convert water to steam. Unlike fuel-fired boilers or electric boilers, an unfired steam generator is an ASME section VIII tank with an internal heat exchanger that does not burn fuel or use electricity as the heat source. In an unfired steam generator, the tube side of the heat exchanger is fed high-pressure steam, thermal fluid, or high-temperature hot water.

Controlling the flow of these heat sources will control the steam output of the steam generator. Unfired steam generators can be horizontal or vertical tanks and are used in a variety of applications today. A facility that has an abundance of any of the three heat sources mentioned previously can be a good application for this type of steam generation. In many areas, steam generation via this method does not require a stationary operator. One particularly useful application of this type of steam generation is pure steam used for sterilization, direct food injection, or drug manufacturing. Unlike fuel-fired boilers, an unfired steam generator can be made from high-grade stainless steel, which can use high-purity water to make very high-purity steam. A popular brand unfired clean steam generator is shown in Figure 7.11.

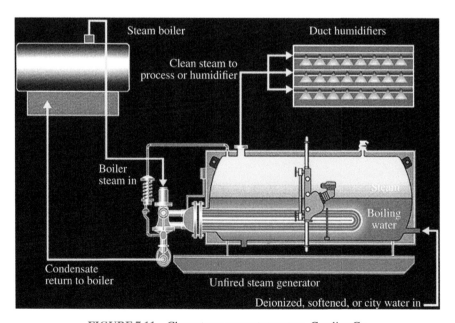

FIGURE 7.11 Clean steam generator system. Cemline Co.

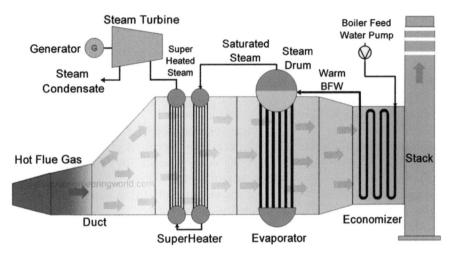

FIGURE 7.12   Typical HRSG Schematic [27].

HEAT RECOVERY STEAM GENERATORS

Heat Recovery Steam Generators or HRSGs are actually a type of unfired steam generator. Instead of using high-pressure steam, thermal fluid or high-temperature hot water as the fuel source, they use waste heat. They are well suited to convert gas turbine exhaust, oven flue gases, or other exhaust gases that are 1000 F or more to generate steam. These are an integral part of a combined cycle cogen plant. HRSGs can be several stories high and contain massive amounts of water. They have level and pressure controls and generally have high surface areas. V. Ganapathy has written a very good book called *Steam Generators and Waste Heat Boilers* [27] that explains in detail the science and engineering behind good HRSG design and application. A typical HRSG sectional design is shown in Figure 7.12. In some instances, a series of gas burners may be added upstream of the superheat section to increase its steam output.

8

BOILER OPERATION AND TRIM

In general, a boiler converts fuel energy into steam. It must therefore have a vessel containing water, a source of fuel, some way of converting the fuel energy into useable thermal energy and control systems to regulate all of these requirements. The center of a boiler operation is an electronic controller sometime referred to as the flame programmer. When the boiler "on" switch is energized, the flame programmer determines if the fuel supply, water level, and steam pressure level are adequate to support steam production. If these conditions are met, then a combustion air fan is started (except for electric boilers) and the controller will determine if there is enough air to support combustion. If this condition is met, then the boiler will perform a series of purges and safety checks to ensure the combustion chamber is clear of tramp fuel vapors. At this point, if everything checks out okay, the flame programmer will allow the pilot fuel valves to open (contacts close for electric boilers), and the boilers burner lights off in pilot mode (usually 2–4%) of rated input. Once the pilot is lit, a flame sensor will sense a flame and provide a small voltage to the flame programmer. If all checks are still okay and an adequate flame is detected, the flame programmer will open the main fuel valves and the boiler will light off on low fire. Once the boiler is lit off in "main" flame mode, the flame programmer will release the burner operation to the steam pressure controls. The controls signal the burner to increase fuel input enough to generate enough steam. Figure 8.12 shows this sequence in a flow chart. For an electric boiler, the sequence is simpler. The boiler controller verifies the need for steam and there is enough water in the pressure vessel. Once these conditions are met, electricity is applied to the electric elements and steam is generated.

Process Steam Systems: A Practical Guide for Operators, Maintainers, Designers, and Educators,
Second Edition. Carey Merritt.
© 2023 John Wiley & Sons, Inc. Published 2023 by John Wiley & Sons, Inc.

THE PACKAGED BOILER CONCEPT

A steam generator is usually purchased and installed as a packaged boiler. A boiler, like a vehicle, can be configured with many options. The key is to purchase the right configuration for the specific application. A simple set up is generally the best; however, some options should always be considered. This chapter will look at each of these options and how they must integrate to provide a packaged steam generating system.

A packaged boiler, shown in Figure 8.1, is an insulated pressure vessel fitted with the water level controls, steam pressure controls, a fuel train, combustion controls, burner, blowdown controls, gauges, and a water level sight glass. Installation of a packaged boiler generally requires only connecting the fuel supply, feed water supply, power source, blowdown drain, combustion air source, and exhaust gas stack. In some cases, the boiler controls may be connected to the building management system or process computer.

Insulated boiler pressure vessel (PV): The core of the steam generator is the pressure vessel. These vessels are ASME code vessels manufactured to the specifications of ASME Section I, Section IV, and Section VIII code. ASME Section I applies to all pressure vessels that are direct fired with a design pressure >15 psig. ASME Section IV applies to all pressure vessels that are direct fired have a design pressure of 15 psig or lower. ASME Section VIII applies to all unfired pressure vessels regardless of the design pressure. Each ASME section has specific requirements for fabrication and testing. Process steam generators will likely be ASME Section I boilers with a **maximum allowable working pressure (MAWP)** greater than 15 psig. All ASME Section I vessels must be designed based on stringent ASME pressure vessel calculations, welded with ASME certified welders, post weld heat

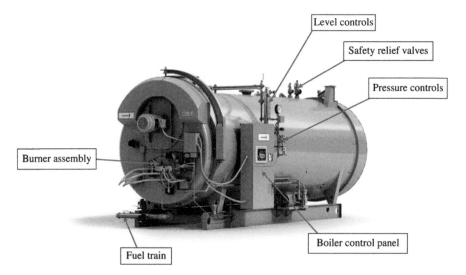

FIGURE 8.1   A packaged boiler. Courtesy of Cleaver-Brooks, Inc.

treated to relieve weld stresses and hydro tested to 1.5 time the vessel MAWP. In addition, most local codes require the **boiler external piping (BEP)** be hydro tested to the same pressure standards as the main pressure vessel. The BEP boundary includes steam stop, feed water, blowdown valves as seen in Figure 13.2. Owners should ensure that they obtain all the **ASME construction and hydro testing documentation (ASME P-Data Forms)**. The pressure vessel is insulated with high-temperature insulation and refractory to minimize radiant heat losses and keep the external surfaces relatively cool. Each pressure vessel will be fitted with a safety relief valve(s) that are sized to relief enough steam at a specified pressure at or below the MAWP. For correct pressure vessel fabrication, the boiler manufacture will need to know what the operating pressure and type of control systems will be used to size and locate the vessel openings correctly.

The water level controls have two distinct purposes. One, they control operating water level in the boiler, and two, they provide a safety shutdown interlock in case of a low water condition. Poor water level control can create an unsafe condition, poor performance, or nuisance shutdowns. There are two types of water level controls, on/off and modulating. On/Off controls will use level sensing devices like a float switches or conductance probes to sense water level. Typically the level sensors sense high water, normal level high, normal level low, primary low water cutoff level, and secondary low water cutoff levels. The **high water level** switch typically drives an alarm only but may also serve to shut down the boiler burner. The **normal water level high** is the level that shuts off a feed water pump or closes a feed water supply valve. The **normal water level low** is the water level that starts a feed water pump or opens a feed water supply valve. The first **low water level cutout** is the water level below the normal water level low that will shut the boiler burner off, but will usually automatically restart the burner once the water level is returned to a level above this first low level cutoff. The second low water cutoff, sometimes called the **auxiliary low water cutoff** is the lowest water level in the boiler the manufacture recommends for safe operation. If this low level is achieved, the normal water level is not being maintained at all and represents an increase in a potential unsafe condition. Consequently, the second low water cutoff level requires manual reset locally by an operator. On/off water level controls are simple and used mainly in smaller steam boilers. The addition of feed water to the boiler using this type of level control tends to make steam production cyclic. The cooler feed water is periodically injected at a rate that equals 1.5–2.5 times the maximum steaming rate. Consequently, when the cooler water is injected, the boiler water is temporarily cooled enough the stop steam production. A diagram showing these types of level control systems are shown in Figure 8.2.

An alternative to on/off level control is modulating level control. Modulating level controls have similar water level switches for high, low, and auxiliary low water conditions. The normal water level low and normal level high represent a band where the level controller will modulate feed water flow to the boiler. Basically the lower the water level in this band, the more flow is allowed to the boiler by opening a flow control valve or increasing the output of a feed water pump. Modulating level control is generally seen on boilers >50 hp. Typical modulating level sensors are float type,

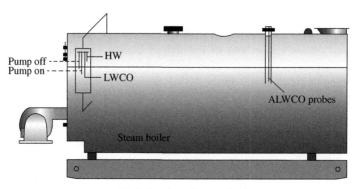

Typical probe style water level sensors

HW—High water
Pump off—Normal water level high
Pump on—Normal water level low
LWCO—Low water cutout
ALWCO—Auxiliary low water cutout

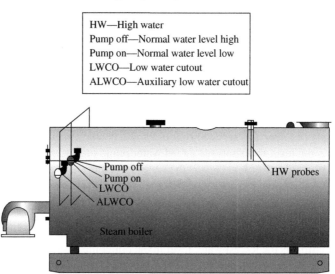

Typical float style water level sensors

FIGURE 8.2   On/off water level control systems.

capacitance type, guided radar, or differential pressure transmitter (DPT) type. Their typical set up is shown below in Figure 8.3. These types of level controls generate an electronic signal proportional to the water level. That electrical signal is used to proportionally open a feed water valve or speed up/slow down a feed water pump motor via a VFD (variable frequency drive) motor. Integration of the level sensor and feed water system are covered in more detail in Chapter 10.

To verify the level control system operates properly, observe the water level over a 30 min time during a period where the boiler is operated at or near the rated pressure and steam flow. The water level should be maintained visible in the site glass between the normal level high and low marks during this period. It may warrant using a marker pen to put marks on the boiler sight glass to show these various levels.

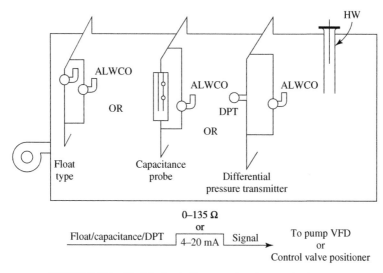

FIGURE 8.3 Modulating boiler water level control schemes.

The pressure controls control the boiler pressure. They are comprised of the operating and high limit pressure sensors/switches. This part of the steam system indirectly controls the amount of steam the boiler generates. Since a boiler does not measure mass flow of steam, it uses steam pressure as a means to regulate output. The pressure sensing equipment senses steam pressure and provides an electronic output signal to the burner controls. This signal tells the burner combustion control system to start, stop, or modulate. Generally there are at least two pressure sensors on the boiler located somewhere in the steam space that control the boiler's operating pressure and provide a high pressure safety interlock. The operating pressure control may be an on/off switch type or modulating operating pressure control type. The high limit switch, similar to the auxiliary low water cutoff switch, is an on/off switch and will require a local manual reset. Typical boiler pressure sensor set up is shown below in Figure 8.4.

The on/off operating pressure switches have a fixed differential set point (magnitude is field adjustable) that starts the burner (or power supply to electric elements for electric boilers) when the pressure falls below the set point plus the differential. Conversely, the same switch stops the burner when the steam pressure rises above the set point plus the differential. The higher the differential, the larger the range of acceptable pressure the boiler tries to achieve. **The high pressure cutoff switch should be set at least 10% higher than the operating switch set point, but never set above the boiler MAWP (maximum allowable working pressure).** A high pressure boiler with a 150 psig MAWP may have an operating pressure switch set at or below 120 psig with a 5 psig differential, and a high pressure cutoff switch set point of 135 psig. This boiler will shut off automatically at 125 psig and restart automatically at 115 psig steam pressure. If at any time the steam pressure rises to 135 psig, the burner will shut down and require manual reset of the high limit pressure switch.

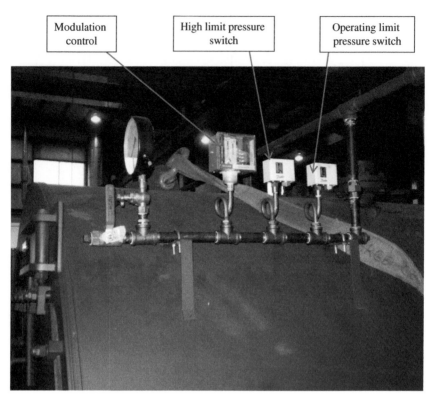

FIGURE 8.4   Installed pressure switches on a boiler pressure vessel [56] Courtesy of the
Yokogawa Electric Corporation [2014].

A modulating operating pressure control system can be a pressure/ohm type, or it
can be a PID (proportional, integral, and differential) type. The pressure/ohm control
senses boiler pressure and converts the pressure into an electronic signal. This signal
is used to position valves on a burner fuel train that control inlet fuel and air. This
relatively simple pressure control has settable differentials and will provide a linear
signal output relative to the boiler pressure. With this type of modulating pressure
control, the differentials are fixed but field adjustable like the on/off pressure
switches.

A PID-type pressure control system uses a pressure sensor mounted in the steam
piping and a digital display mounted where a boiler operator can view and set the
control parameters. The digital controller will have programmable differentials
which allow the operator to set the pressure range with varying differentials above
and below the set point. This feature is very useful when the boiler is significantly
over- or undersized for the system application. Figure 8.5 shows a pressure sensor/
controller setup.

A typical arrangement with a PID-type controller setup for the previously mentioned Figure 8.5 boiler may look like 120 psig operating set point with differentials
of +5 psig and −15 psig. This would give you an operating pressure range of 20 psig.

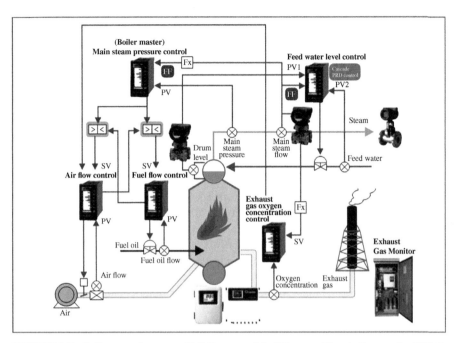

FIGURE 8.5   Boiler control systems. [56] Courtesy of the Yokogawa Electric Corporation [2014].

The burner will shut off at 125 psig but will not restart until the pressure falls below 105 psig. PID-type controllers also have "P"(proportional), "I" (integral), and "D" (differential) settings. The most important setting is the "P" setting which sets the pressure at which the burner starts to modulate. For example, if the "P" setting is 10 and a 0–200 psig pressure sensor is used, then the burner will start to modulate at 0.1 × 200 = 20 psig below the set point. For a 120 psig set point, the burner will start to modulate at 100 psig. Operators and service technicians should read about the PID-type controls and understand how they need to be set up prior to any adjustment of these controls. The boiler equipment manufacture should be able to provide default setting for all three variables.

Regardless of the type of pressure controls installed, the steam system operators must observe the boiler operation through a few cycles and make adjustments that will keep the steam pressure required at the steam using equipment without causing excessive boiler burner cycling.

Blowdown systems: Boiler feed water generally contains some degree of impurities such as suspended and dissolved solids. The impurities can remain and accumulate inside the boiler as the boiler operates. To avoid boiler problems, these dissolved and suspended solids must be periodically discharged or "blown down" from the boiler. The blowdown scheme keeps the boiler water chemistry within specification set forth by the boiler manufacture. Surface blowdown is often done continuously to reduce the level of dissolved solids, and bottom blowdown is performed periodically to remove the suspended solids from the bottom of the boiler. Boiler logs, internal inspections, and water chemistry reports can help monitor the effectiveness of the blowdown practices. Insufficient blowdown may cause carryover of boiler water into the steam or the formation of deposits on boiler tubes. Excessive blowdown wastes energy, water, and treatment chemicals. The amount of blowdown is a function of boiler type, chemical treatment program, feed water quality, and boiler pressure vessel warrantee requirements. Feed water quality is a function of the amount of condensate return. Therefore the more condensate returns, the less boiler blowdown is required.

As previously noted, there are two types of boiler blowdown schemes: surface and bottom blowdowns. The heavier metal corrosion products and precipitates will settle in the bottom of the boiler and boiler operators use a bottom blowdown scheme to periodically purge these solids from the boiler. This method is usually a manual operation performed at least once per day. If only this type of blowdown scheme is used, the operator must check samples frequently and adjust blowdown volume accordingly. This can be labor intensive if the steam system application has a varying load profile or has a low percentage of condensate returns. On small boilers used in closed loop steam systems, the main method of maintaining water chemistry is via manual bottom blowdowns. It is generally better to perform the bottom blowdown after the boiler has set idle for at least one hour. I recommend performing bottom blowdown once per day as part of the daily boiler checks.

Alternatively, surface blowdowns are performed to rid the boiler of dissolved solids. As steam bubbles collapse at the water surface, the steam leaves the boiler and the dissolved solids stay in the water. Consequently, the boiler water near the surface will contain a higher level of dissolved solids. Surface blowdown can be manual or automatic. A manual surface blowdown system uses a set of throttling valves to intermittently or continuously bleed off boiler water. An automatic surface blowdown control constantly monitors boiler water and adjusts the blowdown volume accordingly to maintain the desired water chemistry. Typically, an automatic conductivity-based blowdown uses a probe to measure the boiler water conductivity and provides feedback to a controller which opens and closes a blowdown flow control valve. This type of blowdown control can keep the boiler water chemistry close to the maximum allowable dissolved solids level, while minimizing water discharge and reducing energy losses even during a varying load profile.

A compromise between an automatic conductivity-based blowdown system and a pure manual system is a timed-based automatic blowdown system. In this system, a programmed timer opens a blowdown valve for a specified period. By regulating flow via a throttle valve, frequency of the blowdown and the duration of the blowdown, an operator can maintain the solids level in the boiler. This type of system is ideal for a 8–8 small system with a relatively constant steam load. Examples of boiler blowdown systems are shown in Figure 8.6. I caution the use of the automatic bottom blowdown scheme as you must verify it is allowed in your jurisdiction. I have encountered boiler inspectors and insurance companies that do not allow this configuration.

The last set of boiler trim usually found in a packaged boiler are the **gauges and sight glass**. All steam boilers will come with a steam pressure gauge, but fuel and stack

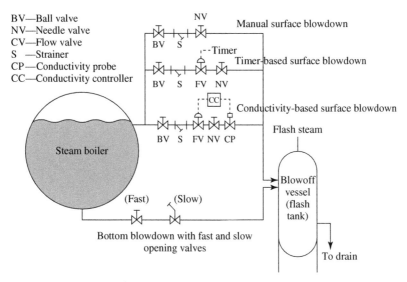

FIGURE 8.6   Boiler blowdown systems.

temperature gauges are often optional. All of these gauges are essential and should be liquid filled, located in an easy to read location and have the right gauge scale. The steam pressure gauge displays the boiler pressure and should be visible from where an operator stands at the boiler control panel. The fuel gauges are really helpful in setting up the high and low fuel pressure switches. The stack thermometer can provide good indication that a boiler is scaled or fouled. Trending stack temperature over a period of weeks will provide a means to determine if good heat transfer is occurring in the boiler. Boiler sight glasses come in a variety of styles and pressure ratings. Boiler manufactures will match the pressure rating of the sight glass to the MAWP of the boiler. A 150 psig sight glass should never be used in a boiler rated for 200 psig.

FUEL DELIVERY AND COMBUSTION SYSTEMS

Fuel trains: The output of a boiler is controlled by the input fuel rate. When the boiler pressure falls below a specified set point, more fuel energy is required to convert water to steam. The fuel train is the part of the boiler that delivers the fuel to the burner. This subsystem includes the gas train or the oil delivery system for a fossil fuel system, a solid fuel feed system for biomass boilers, and electrical contractors or SCR for electric boilers. A typical gas train configuration is shown in the following Figure 8.7. Fuel trains contain a variety of components that control the fuel flow and pressure supply to the burner. Natural gas fuel is generally delivered via underground piping at a relatively consistent pressure. The utility company will usually provide a step-down regulator that will maintain the desired gas pressure from the utility. The boiler gas train will then use another regulator to step the gas pressure down to a level

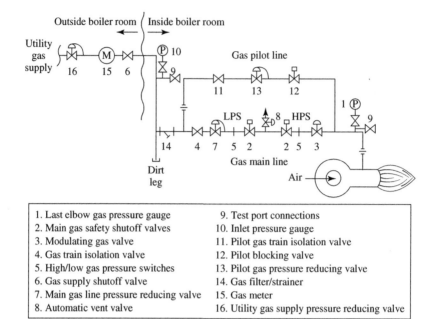

1. Last elbow gas pressure gauge	9. Test port connections
2. Main gas safety shutoff valves	10. Inlet pressure gauge
3. Modulating gas valve	11. Pilot gas train isolation valve
4. Gas train isolation valve	12. Pilot blocking valve
5. High/low gas pressure switches	13. Pilot gas pressure reducing valve
6. Gas supply shutoff valve	14. Gas filter/strainer
7. Main gas line pressure reducing valve	15. Gas meter
8. Automatic vent valve	16. Utility gas supply pressure reducing valve

FIGURE 8.7   Typical double block and bleed gas train configuration.

acceptable for burner operation. System designers must size the gas mains based on allowable line pressure received from the gas company. See Appendix A data to help adequately size gas mains.

Example 8.1

You will be installing two natural gas-fired 20,000 lb/h steam boilers for a new process, and the boilers will require 50,000 scfh of natural gas at 4 psig to the burner. The regulator supplied with the boiler gas train will regulate down to 4 psig with 7–14 psig input pressure. The gas line will need to be run about 1000 feet from the utility regulator to the boiler house. What is the difference in gas line size if the utility provides 10 or 20 psig gas pressure?

*Answer: Since transporting gas will result in a pressure drop, you will need to make sure the gas pressure will be at least 7 psig at the boiler fuel train inlet. Gas line size charts in Appendix A show that a 4 inch line carrying gas 1000 feet will have a capacity of 58,650 scfh but will lose 6% pressure/100 feet of run. A 1000 feet would reduce the supply gas pressure by 60%. Then at 10 psig utility supply pressure, a 4″ line will supply the boiler only 10 psig × 1 – 0.6 = 4 psig. Consequently, **we would need a 6 inch line at 10 psig utility supply pressure to meet the boilers inlet fuel train gas pressure requirement**. At 20 psig, however, a 4 inch line would have a capacity of 56,960 with only a 2% pressure drop per 100 feet. At 1000 feet, the* *pressure at the boilers with a 4" line would then be 20 × 0.8 = 16 psig.* **A 4 inch is large enough at 20 psig utility supply pressure**. *In this application, the 4 inch gas line would require a second pressure regulator to step the line pressure below the 14 psig limit of the burner gas train.*

A boiler burner for an oil-fired boiler will require an oil delivery system [21]. A typical oil delivery system is shown in Figure 8.8. These types of burners will require oil delivered to the burner at a constant pressure, usually 1–2 psig. An oil pump at the burner is then used to increase the oil pressure to >150 psig. The high pressure is required to allow the viscous oil to pass through small nozzles where it can be atomized and burned efficiently. Oil burners will have nozzle(s) which have orifices sized for the right input and have spray angles listed on them for each burner manufacture. The key to burner efficiency is to match the right size orifice and spray angle with the right flow of combustion air. A poorly setup oil burner will not combust cleanly and produce smokey flue gas. The design of the delivery system typically uses a large oil storage tank placed outside which feeds a smaller day tank inside. The day tank allows the oil to reach a constant temperature and provides storage of oil for the boiler(s). Oil burners tends to be a little more difficult to set up and maintain than gas burners. Consequently, increased preventative maintenance and more frequent combustion monitoring should occur with oil-fired burners.

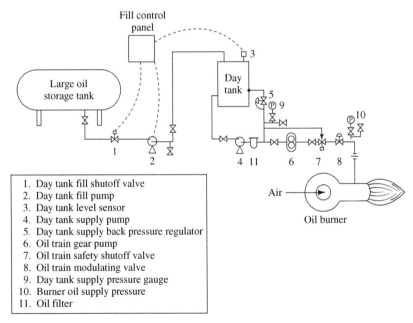

1. Day tank fill shutoff valve
2. Day tank fill pump
3. Day tank level sensor
4. Day tank supply pump
5. Day tank supply back pressure regulator
6. Oil train gear pump
7. Oil train safety shutoff valve
8. Oil train modulating valve
9. Day tank supply pressure gauge
10. Burner oil supply pressure
11. Oil filter

FIGURE 8.8   Typical boiler burner oil delivery system. Data from Preferred Utilities Manufacturing Co. [21].

Regardless of the fossil fuel type, fuel feed will be proportional to the amount of steam needed to be produced. Pressurized gas or oil is supplied to the fuel train and control valves are used to modulate the fuel flow to the burner. In addition to the fuel flow control, simultaneous control of the air flow into the burner is required to ensure proper air/fuel mixture for good combustion. In some systems the air intake dampers are mechanically connected to the gas or oil valves. In this setup, the air automatically modulates with the fuel control valves. Newer combustion controls use control valves equipped with a PLC-type linkage less electronic modulation system to link the fuel and air flow. The PLC-type controls are more complex but generally provide better combustion control over the range of burner inputs. The two different types of controls are shown in Figure 8.9.

A biomass or coal boiler fuel handling system is much different than a traditional fossil fuel burner. The pressure controls typically signal a conveyor or auger system that increases or decreases the feed to the combustion chamber. A typical solid fuel feed system is shown below in Figure 8.10. Solid fuel feed systems generally require a continuously feeding auger system. An operator loads the solid fuel from a stock pile to the feed system. The solid feed moves through the furnace of the boiler where it is combusts. The resulting ash is removed from the furnace via conveyors or fans. The solid fuel feed system is setup for a particular type of feed stock. One important system feature to consider with a biomass or coal boiler is its inability to instantaneously cut

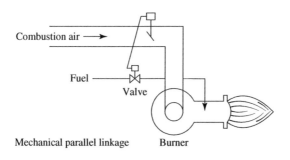

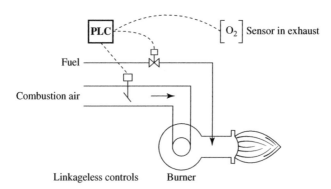

FIGURE 8.9   Burner fuel/air flow control systems.

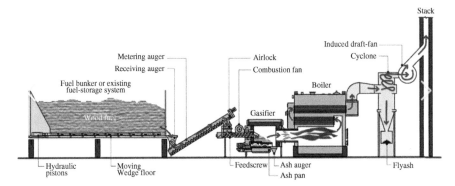

FIGURE 8.10   Wood chip fuel feed system [29].

FIGURE 8.11   Honeywell flame safeguard/programmer (author's collection).

fuel energy input off to the boiler. It will take a few minutes for the biomass, and longer for coal, in the furnace to completely combust before steam production will completely cease. Engineers should always ensure a solid fuel boiler has a means to dump steam when the process load undergoes an emergency shutdown.

Combustion (burner) control: The burners function is to add fuel and air together, and combust them in a safe efficient manner. A device called a **flame programmer** is used to ensure these two functions are accomplished efficiently and safely. Operation of the burner controls can be seen by observing the flame programmer display, burner air, and fuel supply valves during the pre-purge, pilot lighting, main flame lighting, and the burner shutdown operations including postpurge cycles. The flame programmer display will show the status of various phases on the burner light off cycle. The flame programmer gathers data and sends a series of signals to the burner to allow it to complete the combustion cycle. A typical flame programmer is shown in Figure 8.11.

When the burner gets a signal to start the combustion process, the flame programmer will verify the water level in the boiler is above the low water cutoff level, the steam pressure is below the operating set point and signal the combustion air fan to start. If the combustion air flow is adequate (i.e., air switch pulls in) and the fuel pressure to the burner is within the range acceptable for combustion (via high-/low-pressure switch on fuel train). If these parameters are met, then the combustion process is started. However, before any fuel valve is allowed to be opened, all residual fuel vapors must be removed from the entire furnace section. Consequently, the combustion air fan operation is used to perform this task which is called the "prepurge." When the flame programmer initiates the prepurge, the blower motor starts and almost instantaneously the inlet air gate should drive to the full open position and remain there until the prepurge is complete (preprogrammed time from the factory). Once the prepurge is complete, the flame programmer will signal the air gate to drive closed to the low fire position and the flame programmer will then signal the pilot to light. After a few seconds, a flame sensor near the burner head senses the pilot has lit and sends an electronic signal back to the flame programmer. This signal allows the main flame to automatically light on low fire. The flame sensor will sense the main flame has lit and signal the flame programmer to release burner output control to the pressure control system of the boiler.

The burner input is then regulated proportional to how far the boiler pressure is relative to the operating pressure set point. The farther the boiler pressure is from the set point, the higher the rate of fuel is added to the burner. The burner firing rate will then drive toward high fire eventually and modulate at a firing rate that will produce just enough steam to satisfy the demand. This sequence happens for almost all modulating boiler burners. If the burner is an on/off burner, then the purge cycles are performed with a fixed air gate position and the burner main flame will light off on

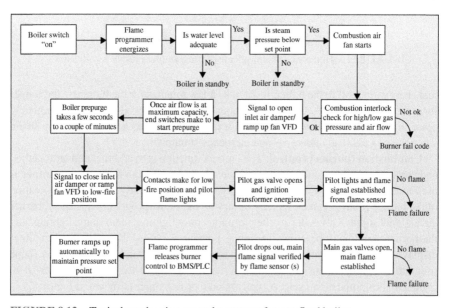

FIGURE 8.12   Typical combustion control sequence for gas-fired boiler.

high fire only. A flow diagram of the various cycles is shown below in Figure 8.12. Typically, on/off burners are used for smaller boilers only.

Burners: Each direct fuel-fired boiler will have some sort of burner attached to the front end of the pressure vessel. The burner's function is to ensure that the fuel and air are mixed together properly to support safe and complete combustion. For liquid and gaseous fuels, the burner will have a flame programmer, as described above, that acts as the "brain" for the burner. Burners come in many sizes and configurations, but most all of them have a control panel that will house the combustion controls and relays required for the safety switches. In addition to the panel, burners will generally come packaged with a fuel train, fan assembly for combustion air, refractory to protect the burner components, flame sensors, combustion air switch, ignition transformer, some means to ignite the fuel mixture and a burner head. Fossil fuel burners can be a gun style or a mesh can type. The gun style injects the fuel through a series of nozzles and the combustion air is added around the nozzles. The fuel/air mixture flows through a spinner assembly that allows for complete mixing. A typical gun-style burner is shown in Figure 8.13.

A mesh-type burner mixes the fuel and air together and ignites it through a circular perforated can assembly. It looks like a large mantle-type assembly. The mesh-style burner head spreads the flame over a greater area and can be mounted closer to the furnace wall. This type of burner is well suited for small compact up to midsize burners wishing to maintain low emissions. A typical mesh-style low emissions-type burner is shown in Figure 8.14.

Oxygen trim: When fuel is combusted with air, it will create carbon dioxide and carbon monoxide. The relative amount of carbon dioxide (CO_2), carbon monoxide (CO), and oxygen (O_2) is proportional to the amount of air added to the fuel. All boiler burners require some excess air to be added to achieve complete combustion. Too little air will produce undesirable carbon monoxide and too much excess air will cause a loss of efficiency. Monitoring CO, O_2, or CO_2 in the combustion flue gases is a good way to ensure the right amount of excess air is being added. Oxygen trim or air density compensation is used to accomplish this

FIGURE 8.13   Gun-style boiler burner. Power Flame Inc.

FIGURE 8.14   Mantle-/mesh-style low NOx boiler burner. Power Flame Inc.

control. Oxygen trim uses an oxygen sensor in the boiler vent piping to sense the oxygen levels in the stack gases. This is very similar to how the oxygen sensors measure engine exhaust oxygen in the exhaust system of our cars. The oxygen sensor detects oxygen levels and converts them into an electronic signal that can be used to position an inlet air damper or control an air supply fan motor VFD. The inlet air damper can be precisely positioned or an air fan motor speed controlled to keep inlet air flow at a level that favors precise combustion. The one downside of oxygen trim is the relatively poor life cycle of the oxygen sensors. They can to fail unexpectedly and periodically over the life of the boiler. When oxygen trim is used, access to these sensors should be included in the boiler house design and spare sensors should be available.

Air density compensation: Another way to optimize combustion is by inlet air density compensation. Air density changes with temperature. Consequently, cold air is more dense than warm air. A fan running at constant speed will move more cold air than hot air. The air flow required to support good combustion, however, is constant at a given firing rate. When combustion air temperature varies, the density varies, and some means to adjust fan speed or air flow will help maintain the right amount of air flow. A temperature sensor can measure the inlet air temperature and, using a PLC, convert the temperature to an electronic signal proportional to its density. An inlet air damper position or air supply fan motor speed can be controlled with this signal creating a fairly precise means to control inlet air to the boiler burner. This type of control is simpler than oxygen trim and is useful when combustion air temperatures vary considerably. If the boiler burner is designed with sealed combustion (outside air source) and significant outside air temperature variations are possible, then O_2 trim or temperature compensation is highly recommended.

LOW EMISSIONS

Low emission burners have become quite popular and in some areas of the country a requirement. Low emissions or LoNOx refers to nitric oxide (NO) and nitrous oxide (NO_2) reduction in the boiler flue gas. Both nitrogen species are represented as NOx and are formed any time fuel is combusted with air. NOx emissions are considered to be significant contributors to air pollution. It has been shown that NOx formationwhen combusting fuel, is a function of nitrogen content in the fuel, combustion flame temperature, or, flame oxygen concentration and the time at that condition [30]. Liquid and solid fuels have higher nitrogen contents and obtaining low emission combustion using front-end technology is difficult. Front-end technology refers to burner technology, where back-end technology refers to postcombustion technology. Both are discussed below.

Research has shown that NOx formation is formed at higher flame temperatures (thermal NOx) and at low oxygen levels (prompt NOx). Figure 8.15 shows the various NOx control technology employed today to lower NOx emissions. Burner manufactures have developed technologies (front end) to allow for good combustion with

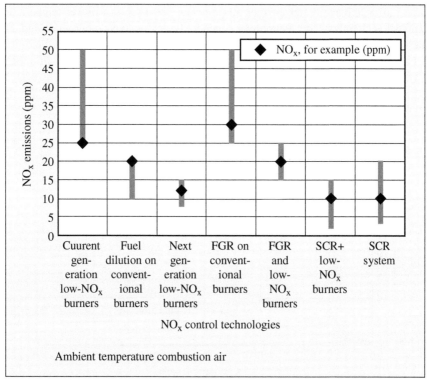

FIGURE 8.15 NOx control technology. Reprinted with permission from *Hydrocarbon Processing* by Gulf Publishing Co. Copyrighted 2001. All rights reserved [31].

reduced NOx formation. Unfortunately, most all of these low emissions burners today can only achieve low NOx levels on gaseous fuel.

Standard gas burners produce NOx levels in the 50–70 ppm range. But several technologies have been used to reduce flue gas NOx levels to <9 ppm. The most widely used low emissions technologies for process-sized boiler systems incorporate flue gas recirculation (FGR), water/steam injection, or staged combustion. All of these technologies effectively lower the flame temperature and reduce NOx formation. Flue gas recirculation injects boiler exhaust gases back into the front of the burner to dilute the flame. FGR essentially reduces the flame temperature and lowers the NOx production. You see FGR used frequently on medium and large boilers. A graphical display of the relationship between NOx formation and FGR levels is shown in Figure 8.16.

Another flame temperature reduction method uses water or steam injection directly into the burner combustion air supply. Although this technology will reduce NOx production, this method will reduce the overall boiler efficiency because the steam or water will rob flame energy. Consequently, this method is a less desirable NOx reduction method than FGR and seen that often.

The last flame temperature reduction technology is called staged combustion. This method injects the fuel at various locations in the burner head to spread the flame out or achieve better air/fuel mixing. The net result is a lower temperature combustion chamber. This technology is generally unique to each burner design and can be used with other NOx reduction technologies. A variation of staged fuel injection is flame manipulation. By changing the shape and formation of the flame using plates and baffles, the flame pattern is modified in a way that increases the surface area and lowers the temperature. Mesh-type burner heads accomplish this by spreading the flame over a larger area.

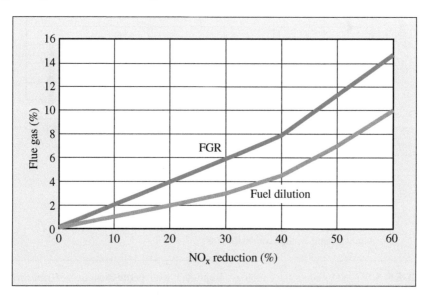

FIGURE 8.16 NOx reduction versus FGR levels and fuel dilution. Reprinted with permission from *Hydrocarbon Processing* by Gulf Publishing Co. Copyrighted 2001. All rights reserved [31].

All of the low emissions technologies mentioned thus far are front-end technology. NOx emission reduction can also be accomplished via back-end technology. These methods treat the flue gas after it leaves the boiler and include selective catalytic reduction and catalytic conversion. Selective catalytic reduction (SCR) uses a reducing agent like ammonia, that is injected into the flue gas where it reacts with the NOx to form water and nitrogen. This technology is used mainly on large industrial and utility boilers.

Catalytic conversion uses a catalyst plated on an inert substrate that when flue gases pass over it, will break down the NOx into nitrogen and oxygen molecules. This technology is similar to the catalytic converter found on automobile engine exhaust. The catalyst cost tends to make this method of NOx reduction quite expensive and is only used in specialized applications [32].

NOx is measured in parts per million (ppm) concentration. It is therefore imperative when comparing burner performance, that NOx concentration in flue gas analyses be corrected to a standard oxygen or carbon dioxide level. Otherwise the NOx level could be misleading when high excess air is used. A standard practice is to correct NOx levels to 3% O_2 in the flue gas [33]. We know from combustion data, that natural gas combusted flue gas will have a CO_2 level of 10.23% at 3% O_2. Therefore the following formula can be used to correct measured flue gas NOx to a concentration of 3% O_2.

$$NOx\ corrected = NOx_{stack}\ measured \times (10.23\% \ /\ measured\ CO_{2stack}\%)$$

Another useful way to express NOx emissions is to determine the mass flow of NOx/unit input fuel. This can be expressed as lb NOx/MMbtu fuel input. This is essentially a mass emission rate per energy of burner input. The mass flow value can be used to calculate annual emissions of NOx for regulatory reporting. The mass flow is calculated by first determining the EF or the Emissions Factor. EF is calculated in lb/MMbtu as per the formula below [34].

$$EF = (C \times F) \times (20.9 \ /\ 20.9 - \%O_2)$$

where 20.9 represents the dry free air oxygen level, C is the concentration of NOx in lb/dscf, and F is the ratio of the gas volume of the products of combustion to the heat content of the fuel minus the water. For natural gas F = 8710 dscf/MMbtu.

In order to determine the EF, we must first convert the measured NOx level from ppm to lb/dscf using the conversion 1 ppm NOx = 0.0000001194 lb/dscf at 0% O_2 [34]. Therefore, a measured NOx concentration of 50 ppm at 3% O_2 would have an EF of

$$EF = (C \times F) \times (20.9 \ /\ 20.9 - 3\%)\ or$$
$$(50 \times 0.0000001194\text{lb/dscf} \times 8710\ \text{dscf/MMbtu}) \times 0.856 = \textbf{0.0445 lb NOx / MMbtu}$$

The resulting NOx mass flow value can be multiplied by the rated boiler input and annual equivalent hours of rated operation to yield the annual weight of NOx discharged to the environment. The annual equivalent hours of operation per year is the annual hours the boiler operates times the average firing rate. Unless you record precise steam metered flow, the average firing rate will be an estimate.

Example 8.2

A 250 hp natural gas steam boiler has an input of 10,780 Mbtu/h and the measured stack NOx level of 50 ppm at 3% O_2. The boiler runs 5 days week 50 weeks per year for 10 h per day at 80% output. What is the annual NOx emission for this boiler?

*Answer: We first have to calculate the EF for this boiler. From the example above, we can see that at 50 ppm NOx at 3% O_2, the EF is 0.0445 lb NOx/MMbtu. The equivalent annual hours of rated input is calculated as (50 wk/yr × 5 days/wk × 10 h/day) × 0.80 firing rate or 2000 h/year equivalent rated output. The annual NOx emissions for this boiler is 0.0445 lb/MMbtu × 10.78 MMbtu/h × 2000 h = **42.4 lb/year**.*

MULTIPLE BOILER SEQUENCING

When more than one boiler is used in parallel, oftentimes, a **Boiler Master Control Panel** will be used. This type of control system is likely to be a PLC-based panel that controls the staging sequence of operation and can display many useful boiler system parameters. Figure 8.17 shows a typical boiler master control panel display.

The master control panel generally has an HMI or human machine interface screen that might display fuel flow, steam flow, feed water temperature, steam pressure, stack temperature, and a host of other useful boiler information. System

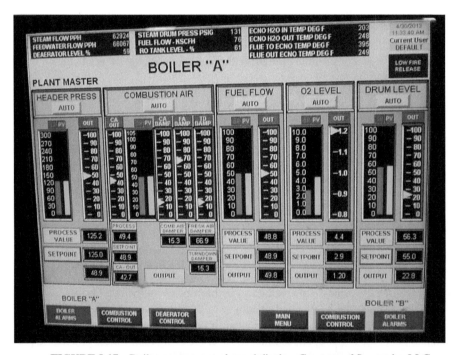

FIGURE 8.17   Boiler master control panel display. Courtesy of Sunmarks, LLC.

operators can view the performance of the boiler room equipment from this panel. Sometimes this panel is located in a main control room far from the boilers themselves. The master boiler control panel usually serves as a lead lag control. This function will alternate which boiler comes on line first and should stage the boilers in manner that allows each boiler to operate in the most efficient output range (generally 25–80% output range). When a process system uses steam from multiple boilers, operations staff should optimize the lead lag sequencing of excessive boiler cycling could occur.

When multiple boilers are mechanically and electrically integrated into a single steam loop, care should be taken to make sure they are staged appropriately. Oftentimes a few different configurations are seen.

a. Boiler redundancy where two or more identical boilers are installed. Often seen in hospitals where the N + 1 requirement is mandatory. This system setup will provide a lot of steam flow flexibility, however, likely be more costing as an initial capital investment.

b. A large and small boiler are installed where the small boiler picks up the shoulder loads when the steam demand is low. The small boiler is sometimes called a "pony" boiler. If the steam demand varies with the seasons (i.e., heating), this configuration is a good setup. Generally both boilers are not service at the same time.

c. A large boiler and small boiler are installed and both operate at the same time. The large boiler operates as a baseline unit and steam flow is modulated with the smaller boiler. If the steam load is high but requires a low tolerance steam pressure variation, this configuration will accomplish this without excessive boiler burner cycling. Combined process and heating loop steam system sometimes use this configuration.

In all cases, when multiple boilers are installed on a single loop, there are some system design good practices that should be followed. Multiple boilers not integrated well can flood or cycle excessively. An idle boiler connected to a common steam loop can flood when not in service. Migrating steam from the steam main will condense and add water to any boiler that is cooler than the steam. Design features such as nonreturn valves, check valves, drip legs, high water equalization lines are means that can be used to protect an idle boiler from flooding. These techniques are shown in Figure 14.5. Nonreturn and check valves physically close to prevent steam migration. Drip legs trap and remove condensed steam before it can get into the idle boiler, and equalization lines keep water level equal in all boiler on the loop. Nonreturn and check valves are used on high pressure boilers. Drip legs and equalization lines are used on low-pressure systems.

It is important to stage boilers so that the overall system is efficient and excessive wear is prevented on any one boiler in the system. Generally, boilers can be staged via the Master Control Panel by simply programming it to meet the system needs. Boilers can be staged based on firing rate or steam pressure. For example, if you have two boilers providing steam to a common loop, then both boilers may carry the load anytime one boiler needs to operate at >75% output. It is generally better (efficiency) to operate two boilers at 50% than one at 100%.

Likewise, steam pressure can be used to stage boilers. If a system requires steam at 110–130 psi pressure, one boiler can have an operating range of 110–120 psig and the other boiler at 110–130 psig. When steam pressure is lower than 120 psig, both boiler are on. As steam pressure goes higher than 120 pasig, one boiler shuts off. By using lead lag programming with pressure differential setting, an operator has tremendous flexibility staging the boilers in the loop. A detailed investigation of the load profile is needed to really fine-tune the best staging strategy.

9

THE STEAM DELIVERY SYSTEM

The steam delivery system is the essential link between the steam generator and the steam-using equipment. An efficient steam delivery system is critical if steam of the right quality and pressure is to be supplied, in the right quantity, to the steam-using equipment. Installation and maintenance of the steam delivery system are important issues, and must be taken into consideration in the design stage. The steam delivery system is comprised of the steam distribution piping, control valves, accumulators, filters, and sensing equipment. Designers need to account for steam line capacity, pressure management, drainage/venting, steam quality, and linear expansion. This chapter will look at each component of the steam delivery system and review proper design and installation criteria.

STEAM FLOW

When the boiler main steam isolating valve (also called the steam stop valve) is opened, steam immediately passes from the boiler into and along the steam mains to any point of lower pressure. The pipework is initially cooler than the steam and some steam on contact with the cooler pipes will begin to condense immediately. Consequently, on startup of the system, the condensing rate will be at its maximum and can actually create a slight vacuum in the header piping. This condensing rate is commonly called the **starting load**. During startup loads, condensation on the cold steam piping dictates the steam flow and can yield very high instantaneous steam demand. The resulting condensation (condensate) falls to the bottom of the pipe and is carried along by the steam flow. **The gradient in the steam main lines should be arranged to slope in the direction of steam flow**. The collected condensate is drained from various strategic points, called drip legs, in the steam main. Once the

Process Steam Systems: A Practical Guide for Operators, Maintainers, Designers, and Educators,
Second Edition. Carey Merritt.
© 2023 John Wiley & Sons, Inc. Published 2023 by John Wiley & Sons, Inc.

steam piping has warmed up, condensation in the main header is minimal (if insulated) and the bulk of the condensation occurs downstream in the steam-using equipment. The steam transfers its energy in warming up the equipment, and, when up to temperature, continues to transfer heat to the product. Control valves can control steam pressure or steam flow. There is now a continuous supply of steam from the boiler to satisfy the connected load. The feed water system modulates water flow into the boiler and the pressure controls modulate fuel delivery to the burner to generate just enough steam to satisfy the steam requirements of the system. At this point the steam process heating requirement in using equipment will dictate actual steam flow. This steam flow rate is commonly called the **"running load."**

STEAM DISTRIBUTION PIPING

The steam delivery system starts with the steam distribution piping. The steam generated in the boiler must be conveyed through piping to the point where its heat energy is required. From the boiler there will be one or more main pipes, or "steam mains," which carry steam from the boiler in the general direction of the steam-using equipment. Smaller branch pipes carry the steam to the individual pieces of equipment. A typical main steam piping diagram is shown below in Figure 9.1.

Correct piping size and orientation are essential for proper operation of the entire steam delivery system. It is beneficial to run high-pressure steam throughout the plant and step the pressure down at each branched location, if required. Various temperatures throughout the process can be achieved by simply regulating the steam pressure at any given location. This ability to manipulate the heat transfer temperature from a common steam supply help make steam an attractive heat transfer fluid. **There are a few rules of thumb that can be applied to steam distribution designs.** Detailed explanations of these rules are contained in the rest of the chapter.

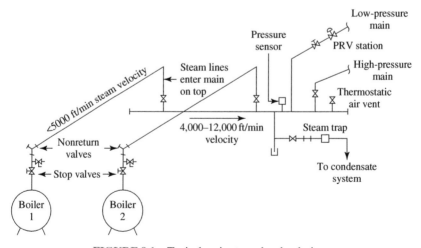

FIGURE 9.1   Typical main steam header design.

- The piping should be sized for steam velocities 6000 ft/min–12,000 ft/min. The lower velocity for steam heating applications and the higher for process applications.
- The piping should have a slight slope gradient toward the flow of steam.
- The drip legs should be sized at ½ the diameter of the main header diameter and 18" minimum drop. Main headers 4" or less diameter should have the same size drip legs.
- The steam line connections from the boiler and to the branch steam lines should always come in the top of the main steam header.
- Restricting orifice plates can be used to retard the flow of steam during startup loads. The orifices must be sized as to not retard flow at rated output.
- Air vents/vacuum breakers should be placed on the steam main header to vent off air and prevent a vacuum causing condensate aspiration back into the steam mains and boiler upon cool down.
- The main steam piping should always be insulated.
- The piping should be sized (for velocity) per the steam operating pressure not the boiler MAWP; however, the steam piping (pressure) should be rated for use that equals or exceeds the boiler's MAWP.
- If pressure reduction is applied, it should be close to the lower pressure steam-using equipment.
- If pressure reduction via a PRV is more than 20 psig, then care must be taken to deal with the resulting superheated steam. See Chapter 14.
- If pressure reduction is used, the steam line between the pressure reducing valve and the steam-using equipment should have a safety relief valve installed to protect the steam-using equipment from over pressurization.
- Expansion loops should be used to reduced pipe stress due to linear expansion from pipe temperature changes.

Steam line draining, venting, and orientation. An essential part of the steam delivery system is the steam line draining, venting, and orientation design. **Failure to properly drain the steam lines will cause water to build up in the lines which can lead to severe water hammer and poor heat transfer**. The correct design is easy to appreciate when you understand what happens in the steam piping from the time steam is first introduced. When a boiler is fired up it starts to produce steam and that steam will migrate into the steam piping. The steam will need to replace the air in the piping; therefore, the air must be vented adequately to allow steam to flow properly. Air vents via **thermostatic air venting** should be placed in the main header and in each branched steam line located close to the steam-using equipment. These thermostatic air vents allow air to pass but will seal steam flow. They also serve a reverse roll when the steam system is shut down. Once steam flow stops, the steam will eventually condense and if not vented properly, will create a vacuum. A vacuum in the steam piping from cooled steam lines is not good and can actually pull water into the boiler from the feed water system. If this occurs, the boiler will likely be in a high water condition when restarted. Oftentimes a vacuum breaker will be installed at the top of a boilers water column piping to prevent a vacuum in the boiler as it cools down.

In an operating steam system, as the steam migrates into the main header and branch piping, it comes into contact with cold metal piping and the steam will condense rapidly until the piping temperature reaches the temperature of the steam. The condensed

water needs to be removed from the steam piping or it will be carried down the piping to the steam-using equipment. Since velocities in the steam piping are quite high, a slug of water can develop very high velocity and cause severe water hammer. The momentum of the water as it slams into the pipe wall where the pipe changes direction, causes a banging sound, and can rip them off their hangers. Therefore it is imperative any water accumulated in the steam piping be removed efficiently. This is done via drip legs as shown in Figure 9.2 below. Notice the dip legs are sized about ½ times the diameter of the header piping size and are at least 18" deep. Steam headers ≤4 inch diameter, however, should have drip legs of equal size as the main header. The drip legs should be installed at least one per 150–200 feet of steam main piping or at any low point condensate dead leg location. Drip legs should also be located upstream of control valves, expansion joints, and pressure reducing stations if steam separators are not used.

All drip legs should be properly trapped and a means to be drained prior to each cold system startup. To estimate the amount of condensate each drip leg will need to remove, use the data in Table 9.1. This table provides the estimated condensate produced in the steam mains carrying steam at various pressures. Note, the data is pounds of condensate per hour per 100 ft of pipe for running loads. Startup loads will be higher. Many process steam system operators will manually open drip leg drains after a cold system startup until the steam mains are fully heated.

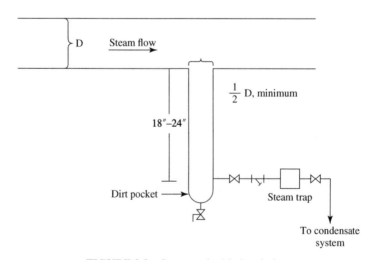

FIGURE 9.2   Steam main drip leg design.

TABLE 9.1   Drip Leg Running Loads (lb of Condensate/h/100 Ft of Steam Main)

Steam Pressure (psig)	Pipe Size														0° Correction Factor*
	2"	2 1/2"	3"	4"	5"	6"	8"	10"	12"	14"	16"	18"	20"	24"	
10	6	7	9	11	13	16	20	24	29	32	36	39	44	53	1.58
30	8	9	11	14	17	20	26	32	38	42	48	51	57	68	1.5
60	10	12	14	18	24	27	33	41	49	54	62	67	74	89	1.45
100	12	15	18	22	28	33	41	51	61	67	77	83	93	111	1.41
125	13	16	20	24	30	36	45	56	68	73	84	90	101	121	1.39
175	16	19	23	26	33	43	53	66	78	86	98	107	119	141	1.38
250	18	22	27	34	42	50	62	77	92	101	116	126	140	168	1.36
300	20	25	30	37	46	54	68	85	101	111	126	138	154	184	1.35
400	23	28	34	43	53	63	80	99	118	130	148	162	180	216	1.33
500	27	33	39	49	61	73	91	114	135	148	170	185	206	246	1.32
600	30	37	44	55	68	82	103	128	152	167	191	208	232	277	1.31
800	36	44	53	69	85	101	131	164	194	214	244	274	305	365	1.3
1000	43	52	63	82	101	120	156	195	231	254	290	326	363	435	1.27
1200	51	62	75	97	119	142	185	230	274	301	343	386	430	515	1.26
1400	60	73	89	114	141	168	219	273	324	356	407	457	509	610	1.25
1600	69	85	103	132	163	195	253	311	375	412	470	528	588	704	1.22
1740	76	93	113	145	179	213	278	347	411	452	516	580	645	773	1.22
1800	79	96	117	150	185	221	288	358	425	467	534	600	667	800	1.21

Data from Watson McDaniel [18].

*For outdoor temperatures of 0°F, multiply load value selected from table by correction factor shown.

Equally important is the orientation of the steam piping. Boiler steam supply lines should be designed to connect to the top of the main header. The header piping and all subsequent **piping should slope a minimum 0.25 inches per 10 feet in the direction of steam flow**.

Pipe sizes: Before steam piping sizing can be done, one should first understand some fundamentals about pipe sizes. There are a number of piping standards in existence around the world, but arguably, the most global are those derived by the American Petroleum Institute (API), where pipes are categorized in schedule numbers. These schedule numbers bear a relation to the pressure and temperature rating of the piping. There are 11 schedules ranging from the lowest at 5 through 10, 20, 30, 40, 60, 80, 100, 120, 140 to schedule No. 160. See Table 9.2 for pipe size data. Schedule 40 is the lightest that would be normally be specified for steam piping applications. Schedule 80 is also used frequently in the steam system. Regardless of schedule number, pipes of a particular size all have the same outside diameter. Consequently, a 2 inch schedule 40 pipe has the same outside diameter as a schedule 80 pipe. As the schedule number increases, the wall thickness increases, and the actual bore is reduced. Therefore, when sizing steam lines, the pipe schedule must be known and the correct inside diameter taken into account. Pipes for steam systems are commonly manufactured from schedule 40 or 80 carbon steel. **The same material may be used for condensate lines, although condensate lines are often schedule 80 carbon steel or schedule 40 stainless steel due to the corrosive nature of low pH condensate**.

Steam line sizing. Choosing the appropriate size piping to carry steam from the boiler to the location where steam will be used is essential. Undersized piping will cause high-pressure drop, velocities, noise, and erosion. Oversized piping creates higher expense to install and insulate, and will have greater heat losses.

Steam piping should be sized to keep the pressure drop or velocity below an acceptable level. See table 9.3 for more guidance. Piping is usually sized based on velocity; however, long runs of piping should also be sized to minimize pressure drop. Consequently, steam velocities are usually kept between 80 and 120 ft/s or 4800 and 7200 ft/min to reduce fiction losses and noise. Some process applications where dry, high-pressure steam is used, can tolerate velocities up to 200 ft/s or 12,000 ft/min. Steam velocities over 8000 ft/min, however, will whistle and should only be used where noise is not a concern. High steam velocities can cause erosion of the steam lines, especially if any water droplets are entrained in the steam. Superheated steam contains no water droplets, and therefore can be sized for higher velocities than saturated steam.

The chart shown below in Figure 9.3 can be used to size steam piping. Find the mass flow of steam desired on the y-axis and follow the horizontal line from left to right until you intersect the steam system operating pressure line. Next, follow that intersection point straight up until it intersects with the desired diagonal steam line size. Lastly, from that intersection point go right horizontally to read the corresponding steam velocity. Notice the chart applies to schedule 40 pipe size. Adjustments must be made for other pipe schedules.

TABLE 9.2 Schedule 40 and 80 Pipe Dimensions

Schedule 40 Pipe Dimensions

Size (inches)	Diameters		Nominal Thickness (inches)	Transverse Areas			Length of Pipe per Square Foot			Cubic Feet per Foot of Pipe	Weight per Foot (pounds)
	External (inches)	Internal (inches)		External (square inches)	Internal (square inches)	Metal (square inches)	External Surface (feet)	Internal Surface (feet)			
1/8	0.84	0.622	0.109	0.84	0.84	0.84	0.84	0.84	0.84	0.84	0.84
3/4	1.05	0.824	0.113	0.866	0.533	0.333	3.637	4.635	0.037	1.13
1	1.315	1.049	0.133	1.358	0.864	0.494	2.904	3.641	0.006	1.678
1 1/4	1.66	1.38	0.14	2.164	1.495	0.669	2.301	2.767	0.01039	2.272
1 1/2	1.900	1.610	0.145	2.835	2.036	0.799	2.010	2.372	0.01414	2.717
2	2.375	2.067	0.154	4.430	3.355	1.075	1.608	1.847	0.02330	3.652
3	3.500	3.068	0.216	9.621	7.393	2.228	1.091	1.245	0.05134	7.575
4	4.500	4.026	0.237	15.90	12.37	3.174	0.848	0.948	0.08840	10.790
6	6.625	6.065	0.280	34.47	28.89	5.581	0.576	0.629	0.2006	18.97
8	8.625	7.981	0.322	58.42	50.02	8.399	0.442	0.478	0.3552	28.55
10	10.750	10.020	0.365	90.76	78.85	11.90	0.355	0.381	0.5476	40.48
12	12.750	11.938	0.406	127.64	111.9	15.74	0.299	0.318	0.7763	53.6
14	14.000	13.125	0.437	153.94	135.3	18.64	0.272	0.280	0.9354	63.0
16	16.000	15.000	0.500	201.50	176.7	24.35	0.238	0.254	1.223	78.0
20	20.000	180814	0.593	314.15	278.0	36.15	0.191	0.203	1.926	123.0
24	24.000	22.626	0.687	452.40	402.10	50.30	0.159	0.169	2.793	171.0

(*Continued*)

TABLE 9.2 (Continued)

Schedule 80 Pipe Dimensions

Size (inches)	Diameters		Nominal Thickness (inches)	Transverse Areas			Length of Pipe (per square foot)		Cubic Feet per Foot of Pipe	Weight per Foot (pounds)
	External (inches)	Internal (inches)		External (square inches)	Internal (square inches)	Metal (square inches)	External Surface (feet)	Internal Surface (feet)		
1/2	0.84	0.546	0.147	0.544	0.234	0.320	4.547	7.000	0.00163	1.00
3/4	1.050	0.742	0.154	0.866	0.433	0.433	3.637	5.15	0.00300	1.47
1	1.315	0.957	0.179	1.358	0.719	0.639	2.904	3.995	0.00500	2.17
1 1/4	1.66	1.38	0.14	2.164	1.495	0.669	2.301	2.767	0.01039	2.272
1 1/2	1.900	1.500	0.200	2.835	1.767	1.068	2.010	2.542	0.01227	3.65
2	2.375	1.939	0.218	4.430	2.953	1.477	1.608	1.970	0.02051	5.02
3	3.500	2.900	0.300	9.621	6.605	3.016	1.091	1.317	0.04587	10.3
4	4.500	3.826	0.337	15.9	11.497	4.407	0.848	0.995	0.0798	14.9
6	6.625	5.761	0.432	34.47	26.067	8.300	0.576	0.673	0.1810	28.6
8	8.625	7.625	0.500	58.42	45.663	12.76	0.442	0.501	0.3171	43.4
10	10.750	9.564	0.593	90.76	71.84	18.92	0.355	0.400	0.4889	64.4
12	12.750	11.376	0.687	127.64	101.64	26.00	0.299	0.336	0.7058	88.6
14	14.000	12.500	0.750	153.64	122.72	31.22	0.272	0.306	0.8522	107.0
16	16.000	14.314	0.843	201.05	160.92	40.13	0.238	0.263	1.117	137.0
20	20.000	17.938	1.031	314.15	252.72	61.43	0.191	0.208	1.755	209.0
24	24.000	21.564	1.218	452.40	365.22	87.18	0.159	0.177	2.536	297.0

Data from Perry's *Chemical Engineering Handbook* [12].

TABLE 9.3 Recommended Steam Velocities and Pressure Drops for Various Services

Service	Velocities	Pressure Drop (per 100 ft)
Saturated Steam		
Vacuum	2000–4000 ft/min	0.25–0.5 psi
0–15 psig	2000–5000 ft/min	0.25–0.5 psi
15–100 psig	2000–7500 ft/min	0.5–1.5 psi
>100 psig	2000–9000 ft/min	0.5–2.0 psi
Steam main, low noise	4000–6000 ft/min	
Steam main, industrial process	8000–12,000 ft/min	
Superheated Steam		
0–100 psig	2500–10,000 ft/min	0.5–1.5 psi
100–800 psig	2500–12,000 ft/min	1.0–2.0 psi

Data from Watson McDaniel [18].

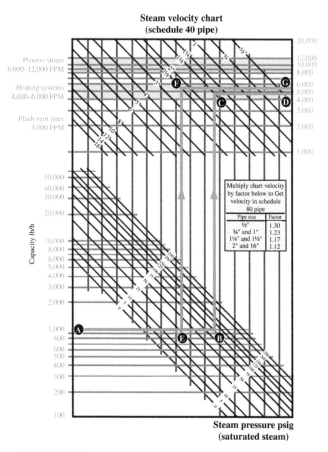

FIGURE 9.3 Steam line sizing chart. Watson McDaniel Co.

The steam velocity in a given size pipe cam also be calculated by the following formula [2].

$$V = \frac{2.4Q\,Vs}{A}$$

where

V = velocity in feet per minute
Q = steam flow in lb/h
Vs = specific volume in cubic feet/lb at the flowing pressure
A = internal area of the pipe in square inches

Example 9.1
You are designing a steam system that will use 20,000 lb/h of 120 psig steam. The process piping will be located close to a control room used to regulate the process. What pipe size and diameter should you run for this application?

Answer: Using the chart in Figure 8.3, we can see that the steam line will be between a 4" and 5" line. When we calculate the velocity in a 4" schedule 40 pipe, we find out the velocity will be V = 2.4Q Vs/A or V = 2.4 × 20,000 lb/h × 3.34 ft³/lb/12.73 sq. in = 12,595 ft/min. Similarly, the velocity in a 6" schedule 40 pipe is calculated at 8016 ft/min. In this application you would choose the 6" main size. Since the steam line is close to a control room, noise from steam flow is a design consideration and steam velocity <8000 ft/min is required. A 6" line is generally preferred because it is a much more common size than a 5" line.

The allowance for pressure losses. As steam passes through the distribution piping, it will inevitably lose pressure due to frictional resistance within the pipework and condensation. Therefore, allowance should be made for this pressure loss when deciding upon the initial distribution pressure. Steam lines that have runs greater than 100 ft should be checked for pressure drop to ensure it is within an acceptable range. The pressure drop in saturated steam distribution pipes can be calculated as per Babcock's equation

$$dp = 0.0001306\,q^2\,l(1 + 3.6/d)/(3600\,\rho\,d^5)$$

where

dp = pressure drop (psi)
q = steam flow rate (lb/h)
l = length of pipe (ft)
d = pipe inside diameter (inches)
ρ = steam density (lb/ft³)

Alternately, the chart shown in Table 9.4 can be used to estimate steam line pressure drop. Acceptable pressure drops will depend on the application. If a process

TABLE 9.4 Sizing Steam Lines Based On Pressure Drop

Pipe Size, Inches (Schedule 40)	lb/h Steam for Piping Pressure Drop of 1 psi/100 ft						lb/h Steam for Piping Pressure Drop of 5 psi/100 ft				
	Steam Pressure (psig)						Steam Pressure (psig)				
	5	10	25	50	100	150	10	25	50	100	150
¾	31	34	43	53	70	84	73	93	120	155	185
1	61	68	86	110	140	170	145	185	235	315	375
1 - ¼	135	150	190	235	310	370	320	410	520	690	820
1 - ½	210	230	290	370	485	570	500	640	810	1050	1300
2	425	470	590	750	980	1150	1000	1300	1650	2150	2600
2- ½	700	780	980	1250	1600	1900	1650	2150	2700	3600	4250
3	1280	1450	1800	2250	2950	3500	3050	3900	4300	6600	7800
4	2700	3000	3800	4750	6200	7400	6500	8200	10500	14000	16500
6	8200	9200	11500	14500	19000	22500	19500	25000	31500	42000	50000
8	17000	19000	24000	30000	39500	47000	41000	52000	66000	88000	105000

These flows were calculated from Babcock's equation.
Data from Eclipse Inc. [35].

requires 120 psig steam far from the point of steam generation, and the boilers are producing steam at 125 psig, then only a 5 psig system pressure drop is acceptable.

Example 9.2
A 4" schedule 40 carbon steel steam main carrying 11,000 lb/h of 100 psig steam has a total linear foot length of 1085 ft. What is the pressure drop at the end of the run?

*Answer: To find the pressure drop, you will need to first find the pipe inside diameter and steam density at 100 psig. From Table 9.2, the inside diameter of 4" schedule 40 pipe is 4.036 inches and from steam charts, steam at 100 psig has a density of 0.26 lb/ft³. The pressure drop is calculated as $dp = 0.0001306\ q^2\ l\ (1 + 3.6/d)/(3600\ \rho\ d^5)$. So $dp = 0.0001306 \times 11,000^2\ lb/h \times 1085\ ft \times (1 + 3.6/4.026\ in)/3600 \times 0.26\ lb/ft^3 \times 4.026^5 = $ **32.7 psig.** This example shows a significant pressure drop; therefore, you would upsize the piping to 6". Incidentally, at 6" schedule 40 carbon steel pipe, the total pressure drop becomes only **3.6 psig.***

Steam at higher pressure occupies less volume than at a lower pressure. It follows that, if steam is generated in the boiler at a high pressure and also distributed at a high pressure, the size of the distribution mains will be smaller than that for a low-pressure system for the same heat load. Generating and distributing steam at higher pressure offers three important advantages:

- The thermal storage capacity of the boiler is increased, helping it to cope more efficiently with fluctuating loads, minimizing the risk of producing poor steam quality.
- Smaller bore steam mains result in lower capital cost for materials such as pipes, flanges, supports, and labor.
- Smaller bore steam mains cost less to insulate.

It may be necessary to reduce the steam pressure to each zone or point of use in the system in order to correspond with the maximum pressure allowed by the application. Local pressure reduction to suit individual plant equipment will also result in drier steam at the point of use. Line size can be a big issue if steam lines are sized for high pressure and steam system operating pressure is lowered significantly. This is the case many times in a retrofit application where the steam boiler operating pressure is lowered to improve boiler efficiency or to avert stationary operator laws. **Retrofitting high-pressure steam systems to operate with low-pressure steam will require evaluation of the steam piping size**.

Example 9.3
A 2000 lb/h steam system operated originally at 100 psig would require a line size of 2"; however, it would require a 4" line at 10 psig. Trying to operate this 100 psig designed steam system at 10 psig with the same mass flow using a 2" line would yield

steam velocities in excess of 12000 ft/min. The higher velocity would have higher pressure losses and it might create more noise. Both could be of concern for the application.

The allowance for the heat losses from the steam piping. When steam flows from the boiler, it starts to give up energy to the metal in the piping. Therefore, we must account for this energy loss to ensure the adequate amount of steam is allowed to reach the steam-using equipment. As the size of the main is yet to be determined, the true calculations cannot be made, but, assuming that the main is insulated, **it is reasonable to add 2% of the steam load per 1000 ft of the length to account for heat losses**.

Sizing pipes for superheated steam duty. Superheated steam can be considered as a dry gas and therefore carries no moisture. Consequently, there is no chance of pipe erosion due to suspended water droplets, and steam velocities can be as high as 20,000 ft/min if the pressure drop losses permit this. This high velocity will create steam flow noise and must be taken into consideration in the placement of steam mains.

Allowances for pipe expansion. Steam lines like any long piece of metal will expand and contract with the addition or removal of heat. Design engineers should take thermal expansion into account if steam lines are being run over a long distance. Many process industries are set up such that the boiler house is away from the process and steam lines are run along pipe racks for hundreds of feet. If pipe movement is not accounted for in the pipe rack and expansion loops are not designed into the long runs, excessive pipe stress will occur. Excessive pipe stress can lead to pipe rack distortions, sprung lines, and weld failures. Pipe expansion can be calculated using the formula [16]:

$$\text{Expansion}\left(\Delta\right) = L_o \times \Delta T \times \alpha$$

where

L_o = length of pipe between anchors (ft)
ΔT = temperature difference (F)
α = expansion coefficient

Table 9.5 below can be used as general guide to determine the expansion coefficient; however, every type of metallurgy has its own expansion coefficient and can be found in the literature. Designers should consult the literature if the piping metallurgy they wish to use is not shown in this table.

Once pipe expansion is determined, then the stresses from the expansion must be mitigated. This is done by adding expansion loops and pipe guide supports that allow movement between pipe anchor points. Figure 9.4 below shows typical expansion loop types, pipe loops, and bellows type. There are two ways to determine where and how many and the size of each expansion loops are necessary. One, run a software program by a credible engineering service like as found on the **www.Engineering ToolBox. com** website under *Steel Pipes: Calculating Thermal Expansion Loops*, [36] or calculate the loop configuration based on the formula (also found on this website [5])

TABLE 9.5   Expansion Coefficient (α)

For Temp Range (°F)	30 to 32	32 to 212	32 to 400	32 to 600	32 to 750	32 to 900	32 to 1100	32 to 1300
Mild steel 0.1–0.2% C	7.1	7.8	8.3	8.7	9.0	9.5	9.7	—
Alloy steel 1% CR 1/3% Mo	7.7	8.0	8.4	8.8	9.2	9.6	9.8	—
Stainless steel 1% CR 1/3% Mo	10.8	11.1	11.5	11.8	12.1	12.4	12.6	12.8

Expansion coefficient $\alpha \times 10^{-5}$ (inches)
Example $7{\cdot}1 \times 10^{-5} = 0.000071$

Data from Spirax Sarco [16].

$$w = 0.015 \left(dl\, D \right)^{1/2}$$

where

w = width of loop (in)
$dl = \alpha$ dt S = temperature expansion (ft)
α = temperature expansion coefficient (10^{-6} in/in-F)
dt = temperature change (°F)
S = length of expanding pipe (ft)
D = diameter of pipe (in)

The offset of the loop is $2\,w$. The length of the loop is $5\,w$.

A diagram of a typical expansion loop showing relative dimensions is shown in Figure 9.5. **Some rules of thumb apply to pipe type expansion loop design.**

- The expansion loop is generally located in the hottest line
- The expansion loop should be located close to the center distance between the anchor points

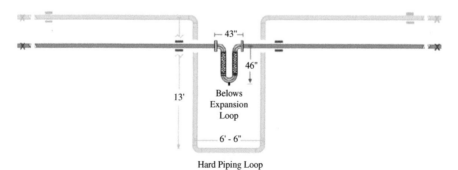

FIGURE 9.4   Typical expansion loop configuration in main steam line.

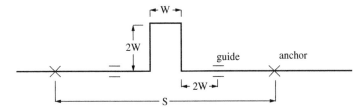

Where S = length of expanded steam pipe

FIGURE 9.5   Steam line expansion loop (nonbellow type) relative dimensions [36].

• The height of the expansion loop is generally at least 2X the width.
• The expansion loop should be run horizontally if possible.

CONTROL VALVES

Steam control valves are divided into two categories, pressure control and temperature control. Since steam temperature is related to its pressure, this concept can be
confusing. I have found the best way to understand the difference is to look at what
each control mechanism is actually trying to maintain. Pressure control valves control steam line pressure to a level that will limit the down steam equipment to prevent
from over pressurization. For example a heat exchanger designed to heat a process
liquid using high-pressure steam may only be rated for 50 psig maximum steam
pressure. A pressure control valve (PCV) will ensure the downstream pressure not
exceed 50 psig regardless of the supply steam pressure. Alternatively, a process
design may wish to limit the steam pressure to 50 psig because of product temperature limitations or wish to keep the pressure gradient across the heat in such a manner
that will only affect process liquid flow one way if there was a tube leak. Again, a
PCV is often used to keep the steam pressure below a set process equipment limit.

Temperature control is used to limit the steam flow to maintain process or product
temperature that is heated by the steam. Where PCVs are controlled by what is happening in the steam line, temperature control valves (TCV) are controlled by what is
happening in the process or product line. The best illustration of this is to look at a
temperature control of a steam to hot water heat exchanger. The TCV allows an
amount of steam to enter the heat exchanger just enough to maintain the flow of hot
water at a specified temperature exiting the heat exchanger.

It is not too hard to see that TCVs and PCVs can be the same type valve but using different sensors located in different places. In some applications both a TCV and PCV is
incorporated into the design. Figure 9.6 shows an application where both TCV and PCV's
are used in a process application to heat up a slurry using steam as the heat source.

Types of control valves. Control valve selection is as important as pipe sizing, but
sometimes misapplied. These valves control steam flow to maintain a desired temperature or pressure. Control valves are usually direct acting, pilot acting, or pneumatic
operated. The cost and degree of control for these valves vary tremendously.

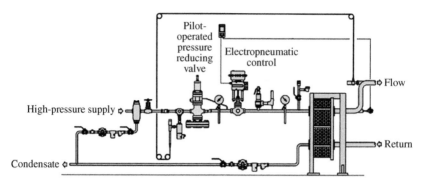

FIGURE 9.6 Pressure and temperature control valves used together. Spirax Sarco Inc.

Direct acting valves are the simplest and least expensive; however, they have relatively low capacity for their size, react relatively slowly, and will regulate with less precision than the other two types of control valves. A typical direct acting valve is shown in figure 9.7. Direct acting valves use a mechanically linked sensor filled with a liquid that expands and contracts with temperature. The sensing line (mechanical link) opens or closes a diaphragm in the valve body allowing steam flow regulation. Consequently, this sensor, when placed in a product provides a means to open or close the valve as the temperature approaches or falls away from a preset set point on the valve. Reaction time can be up to several seconds, so temperature fluctuations in the product may result. The simplicity, low cost make this valve type selection perfectly acceptable for any process that does not require precise temperature control. An example is a water tank that is being preheated with steam or a heat exchanger making hot water with steam where slight hot water temperature fluctuations are acceptable. Sometimes a direct acting valve is used with a pilot operated valve as shown in Figure 9.8.

For precise steam flow regulation, as seen in many process applications, a **pneumatic operated steam control valve** will provide the most precise control. Unfortunately, this type is the most costly. Pneumatically operated control valves use actuators and positioners piloted by controllers to maintain precise steam flow control. The controllers use sensors that sense downstream temperature or pressure fluctuations, interpolate the signals and regulate air supply to a pneumatic positioner which then supplies air to a diaphragm opening or closing a valve. Minute pressure or temperature fluctuations can be sensed with modern sensors that convert steam pressure or product temperature into an electronic signal to the controller. Springs are used to compensate for the loss of air supply and force the valve to fully open (fail open type) or fully close (fail close type). The rate of closure (and opening) can also be controlled by regulating the air supply to the actuator. Many processes used today require very tight tolerances and a pneumatically operated control valve is the valve of choice. A typical pneumatic actuated control valve is shown below in Figure 9.9.

A good compromise between a direct acting valve and a pneumatic actuated valve is a **pilot-operated valve**. These valves will have response times between a pneumatic and direct acting valve and are reasonably priced. Accuracy is good and different

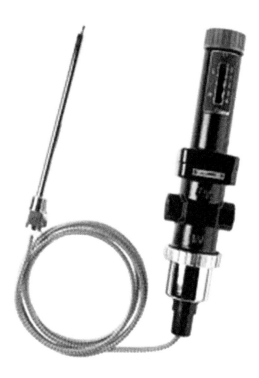

Color-coded handles (red for heating and blue for cooling) make for easy identification in the field.

Easy no-tools temperature adjustment with a simple turn of the handle

Cast bronze body permits liquid service to 250 psi and steam service to 150 psi

Main valve seat materials are stainless steel and Teflon for high durability and positive sealing.

Quick installation and removal of sensor from main body mean easy temperature range changes.

Sensor temperature endurance- +72°F of the maximum value of the temperature regulation range, Gas-charged capillary eliminates change of capillary charge mixing with process in case ofbreakage. Sensor can be mounted in any position.

Sensors are standard for both heating and cooling for all sizes (½″–1″), standard capillary accurate to within ±7°F

A single valve with bellows and balancing mechanism ensures stable regulation. Not affected by pressure fluctuation.

Braided stainless steel capillary protects against crimping.

FIGURE 9.7 Direct acting control valves. [2014] Reproduced with Permission from Armstrong International, Inc.

Control of temperature for storage tanks

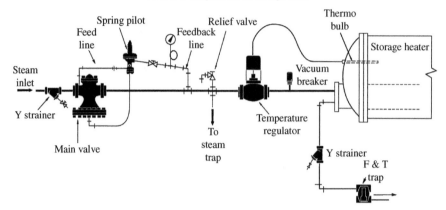

FIGURE 9.8 Direct acting control valve application with pressure regulating pilot operated valve. Courtesy of Xylem Inc.

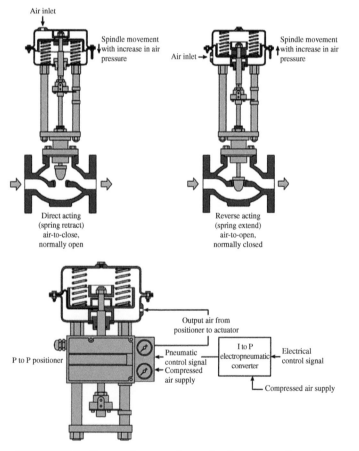

FIGURE 9.9 Pneumatic actuated control valves. Spirax Sarco Inc.

types of pilots (i.e., pressure or temperature) can be used. A pilot operated valve with variuos pilots is shown in figure 9.10. Another advantage of the pilot operated valves is that the valves can have more than one pilot. Sometimes it is beneficial to have a valve control either pressure or temperature and be configured with a shut off function. This is accomplished by using a pressure or temperature pilot and a solenoid shut off pilot. Generally speaking, pilot-operated valves are the valve of choice for pressure control and pneumatic actuated valves are the valve of choice for temperature control.

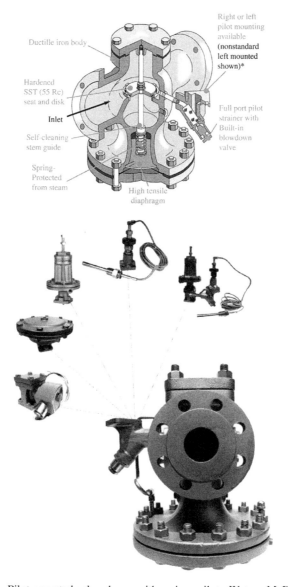

FIGURE 9.10   Pilot-operated valve shown with various pilots. Watson McDaniel Co.

An application requiring a pilot-operated valve to control pressure and a pneumatic actuated valve to control temperature is shown in Figure 9.6.

Opening and closing speed. Steam system designers must ensure steam control valves have slow opening times or rapid boiler depressurization can occur. When a steam control valve opens quickly, the steam pressure in the supply line falls rapidly and can cause boiler pressure to fall radically. Consequently, boiling in the boiler becomes erratic due to rapid bubble size increase. The net consequence is a rapid growth in the boiler water surface foam layer which can result carryover. Opening of control valves too fast can create wet steam and rapid water level drop in the boiler. Pneumatic control valves are subject to rapid response and must be adjusted to regulate valve stroke times. Likewise there are consequences with fast closing valves. Water hammer in feed water, condensate and makeup water lines can result if control valves close rapidly. To remedy water hammer, slow down the closure of these valves or install surge suppressors as described in the section on water hammer.

Pressure reducing stations. The common method for reducing steam pressure is to use a pressure reducing valve, similar to the one shown in the pressure reducing station in Figure 9.11 below.

Notice a separator is installed upstream of the reducing valve to remove entrained water from incoming wet steam, thereby ensuring high-quality steam to pass through the reducing valve. Plant equipment downstream of the pressure reducing valve is protected by a safety valve. If the pressure reducing valve fails, the downstream pressure may rise above the maximum allowable working pressure of the steam-using equipment. This, in turn, may permanently damage the equipment, and, more importantly, constitute a danger to personnel. With a safety valve fitted, any excess pressure is vented through the valve and will prevent overpressurization. A typical internally piloted pressure reducing valve is shown in Figure 9.12

Other components included in the pressure reducing valve station are

- The primary isolating valve: To shut the system down for maintenance.
- The primary pressure gauge: To monitor the supply steam pressure.
- The strainer: To keep the PRV internals clean.
- The secondary pressure gauge: To set and monitor the downstream pressure.
- The secondary isolating valve: To assist in setting the downstream pressure on no-load conditions.

System designers should specify a minimum 10 pipe diameters upstream and 20 pipe diameters downstream of clear piping on either side of pressure regulating valve to ensure only laminar flow through the control valve.

STEAM ACCUMULATION

During certain times of a process steam cycle, the steam demand can spike and cause a rapid boiler depressurization. If a boiler is too rapidly depressurized, the violent resultant boiling can cause boiler water to become entrained in the steam spray and

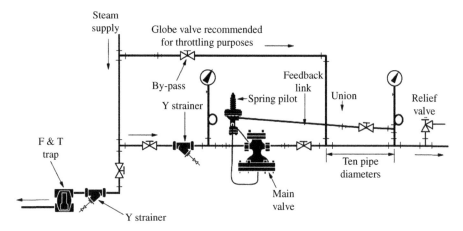

FIGURE 9.11 Typical pressure reducing station. Courtesy of Xylem Inc.

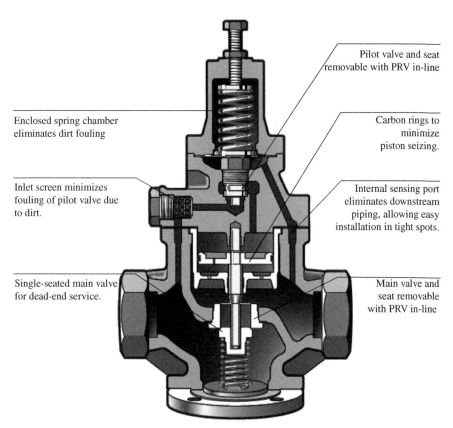

FIGURE 9.12 Internally piloted pressure reducing valve. [2014] Reproduced with Permission from Armstrong International, Inc.

carried over into the steam piping. To help mitigate high steam demand spikes, system designers can either oversize the boiler, design in back pressure regulators to control depressurization, or use a steam accumulator to mitigate the consequences of high instantaneous steam demand.

Facility managers generally will not bear the additional cost of an oversized boiler. The use of a back pressure regulator or an orifice plate in the steam line between the boiler and the steam-using equipment provides a restriction in the steam line to retard the steam flow during high demand periods. This will work for small steam flow spikes or very short period spikes; however, the steam-using equipment may be starved of steam during normal demand periods. Furthermore, a back pressure regulator or restricting orifice adds more unwanted pressure drop in the steam piping.

One solution to this challenge is to incorporate steam accumulation equipment into the steam system design. The steam flow graph shown below in Figure 9.13 demonstrates the load profile from a typical autoclave and the steam boiler system servicing the autoclave without a steam accumulator. An autoclave will have high steam demand at the beginning of a heating cycle. The graph shows if steam accumulation is not used, cycle times for the autoclave will be longer than desired. Essentially, the autoclave will be starved during the time when steam demand is at its highest level.

Two design enhancements can be incorporated to minimize the consequences of high instantaneous steam load demands. The first enhancement is the use of a dry steam accumulator and the second is the use of a wet steam accumulator.

Both design enhancements increase the mass of stored steam. Either type of steam accumulation will not create steam; rather, they are a means to store steam that is available during bursts of high steam demand. Only adding fuel energy to a boiler will

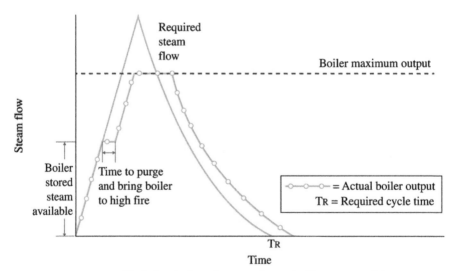

Typical autoclave steam requirement with no accumulator
and standard prepurge flame establishment

FIGURE 9.13   Autoclave steam demand without a steam accumulator.

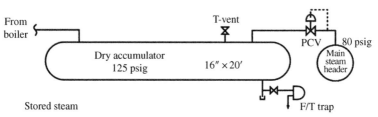

Dry accumulator

1. Steam @ 125 psig has a specific volume of 3.23 ft/lb.

   - The 16″ × 20′ dry accumulator has a volume of 25.6 ft³.

   - The stored steam is then 25.6/3.23 = 7.9 lb of steam.

FIGURE 9.14   A dry accumulator.

create more steam. Dry accumulators are actually an oversized section of pipe or a dry steam storage tank. Simply adding this extra steam space adds extra volume to store steam from the boiler steam space. The extra steam will slow down the depressurization rate of the boiler and help mitigate water carryover from the boiler. When a dry accumulator is used with a pressure control valve as shown in Figure 9.14 above, it becomes effective in mitigating the effects of high instantaneous steam demand. Other advantages of a dry accumulator include boiler protection from excessive cycling. The added steam storage will require extended boiler operation times to recharge the added volume of steam. An accumulator acts like a storage battery of steam.

Example 9.4

A 125 psig steam system with 100 ft of 1.5 inch diameter steam piping needs added steam storage because the boiler is shutting down on low water level during a spike in steam demand. Will adding a dry accumulator help?

Answer: We can increase the stored steam mass significantly if we replace 20 feet of 1.5 inch pipe with a 16″ diameter section of pipe. This oversize section of pipe will be the "dry accumulator." A 16″ × 20″ of pipe has a volume of 25.6 cubic ft. At 125 psig and a steam specific volume of 3.23 cubic ft/lb, this equates to almost 8 lb of steam storage. By adding this dry accumulator, we can increase the dry steam storage more than fivefold. Figure 9.14 shows the dry accumulator for this example.

If more instantaneous steam is required than a dry accumulator can supply, then a wet accumulator can be used. A wet accumulator is a pressurized vessel with hot water connected to the boiler steam line. This vessel is pressurized to the boiler operating pressure and will discharge stored flash steam when the header is depressurized. Once depressurized, the boiler will recharge the accumulator when the load equipment no longer requires steam. Therefore, during idle periods of the steam use cycle, the accumulator can be fully recharged and be readied for the next cycle. The

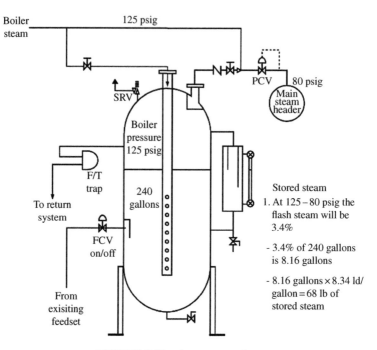

FIGURE 9.15   A wet accumulator.

amount of stored steam is proportional to the water volume and the change in pressure. Using flash steam charts, similar to the one shown in Table 4.1, one can calculate how much flash steam will be available as the steam line pressure starts to decrease. The wet accumulator shown in Figure 9.15 above shows how much stored steam will be made instantaneously available with a wet accumulator holding 240 gallons at 125 psig that is depressurized to 80 psig. It is quite easy to see that a wet accumulator will store significantly more steam than a similar size dry accumulator. Interestingly, the boiler pressure vessel with normal water level is actually a wet accumulator. Applying the use of steam accumulators to the original autoclave steam flow diagram shown in Figure 9.13 will result in shortened cycle times. This is shown in Figure 9.16 below.

Sized right, steam accumulators will greatly shorten cycle times of autoclaves and similar equipment that have spikes in steam demand while maintaining good steam quality.

STEAM FILTRATION

Steam can carry with it unwanted entrained water, dissolved solids, and particulates. All of these contaminates are detrimental to steam piping, control valves, sensors, and steam-using equipment. One means to help maintain high-quality steam is to use steam filtration. Steam filtration can be accomplished using three different types of equipment: strainers, separators, and filters. **Steam strainers** are similar in design to the water strainers used in water lines. They force the steam to flow through a wire mesh and are useful for removing particulates only. Steam strainers are used in upstream of control valves, steam traps, and sensors to remove relatively large (>50 micron) particulate matter. **Steam separators** use centrifugal force to separate the entrained water from the steam. A typical steam separator internal design is shown below in Figure 9.17. They are simple and are quite effective at lowering the entrained water level to <.5%. Consequently, the removal of the water will also remove dissolved and particulate matter. The pressure drop across a steam separator should be less than 2 psig but is dependent on steam velocity and mass flow through the unit.

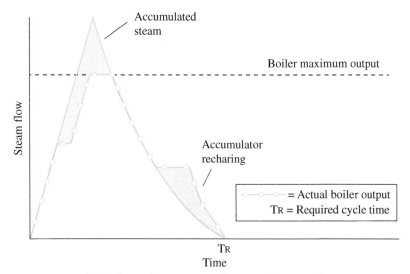

Typical autoclave steam requirement with accumulator
and standard prepurge flame establishment

FIGURE 9.16   Autoclave steam flow using a steam accumulator.

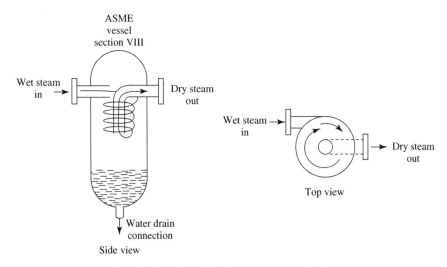

FIGURE 9.17   Typical steam separator design.

FIGURE 9.18   Steam filter. Microdyne Products Corp.

Steam separators will remove almost all of the entrained contaminates; however, some process applications require pure steam. Those applications will require the use of **steam filters**. Direct injection of steam into food products during processing generally will require food grade (3A) quality steam and therefore require steam filtration. Steam filters remove all entrained water and particulates from the steam by forcing the steam through a sintered metal or other coalescing-type filter media. Steam filtering creates higher pressure drops and generally costs are significantly greater than steam separation. Figure 9.18 shows an example of a steam filter.

SENSORS AND METERS

In the steam delivery piping there will likely be sensors. Sensors include pressure sensors, steam flow meters, temperature sensors, and gauges. Also included in this category would be thermostatic vents and vacuum breakers. Sensors should be located on the top of the steam piping and oriented to facilitate easy removal and calibration. A good steam delivery system design will incorporate extra openings for future sensors. Installers need to ensure sensor electrical wiring is protected from the radiant heat from the steam line itself. Remember to locate sensors at least 10 pipe diameters down stream of a valve, elbow, or any change in pipe diameter to ensure adequate mixing and laminar flow.

Steam Metering

A useful boiler output measurement is steam flow. Steam flow can be done anywhere in the steam delivery piping; however, steam flow is normally measured in a horizontal header section where laminar flow exists. There are several types of steam meters, but vortex, turbine, and differential pressure types are the most common. A good review of each type and their applications can be found on Spirax Sarco's website. Turbine meters perform well at low flow rates, are relative inexpensive, but require an additional temperature probe to determine mass flow. Differential pressure type steam meters are quite accurate if the orifice plate is sized correctly; however, these are more costly to install and usually require more maintenance. This type of meter uses a multivariable DP transmitter to measure pressure and temperature. I have found steam vortex meters the easiest and most reliable to use. They come in multiple sizes and can provide steam temperature, pressure, and dryness fraction in addition to steam flow. Care should be taken to ensure the right size vortex meter is selected as many types have threshold steam flow levels before metering commences.

Accurate steam metering can be difficult if the steam flow varies over a wide range. In such cases, it may better to use condensate flow back to the feed water system as a means to measure steam output. Obviously, the steam system must be a closed system and steam leakage minimized to use this method of determining steam output. A typical vortex steam meter installation is shown below in Figure 9.19.

STOP AND SAFETY VALVES

Stop valves are the valves used to isolate steam equipment. In a boiler design most manufactures supply a stop valve located close to the boiler's steam outlet. Steam stop valves are usually gate valves with a rising stem design. Small steam lines may also use ball valves as isolation valves. Globe valves are generally used to throttle steam or water lines (i.e., feed water flow). Globe and gate valve design are shown in Figure 9.20.

A specialized type of stop valve is called the stop check or nonreturn valve. This valve is both a stop valve and a check valve. They can be straight or angle pattern design. Stop check valves have to be sized appropriately or the valve will chatter is low flow conditions. Chatter is actually the check valve oscillating open and close. Care should be taken to make sure these type of valves are not sized too large. Sometimes the stop check valve will be a smaller size than the steam piping they get mounted in. Figure 9.21 shows a typical stop check valve cross section. Nonreturn

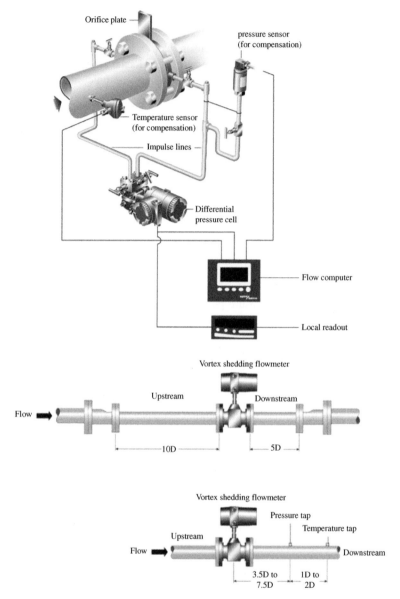

FIGURE 9.19 Typical orifice plate and vortex steam meter design. Courtesy of Spirax Sarco [37].

valve applications are limited to high-pressure steam. Low-pressure steam flow causes chattering.

The last type of valve that is found in a steam system is a safety relief valve or SRV. A typical SRV cut away is shown in Figure 9.22. These valves are relief valves

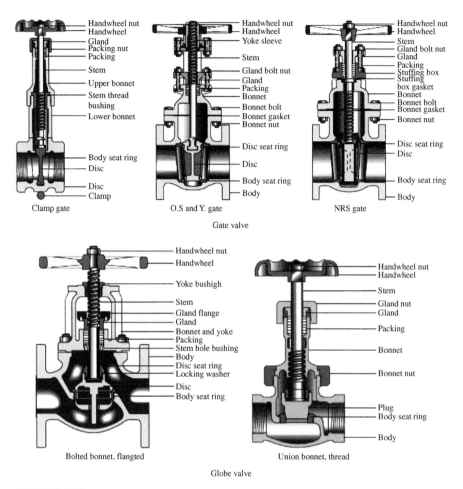

Handwheel nut
Handwheel
Gland
Packing nut
Packing
Stem
Upper bonnet
Stem thread
bushing
Lower bonnet

Body seat ring
Disc

Disc
Clamp

Clamp gate

Handwheel nut
Handwheel
Yoke sleeve

Stem

Gland bolt nut
Gland
Packing
Bonnet
Bonnet bolt
Bonnet gasket
Bonnet nut

Disc seat ring

Disc

Body seat ring

Body

O.S and Y. gate

Handwheel nut
Handwheel
Stem
Gland bolt nut
Gland
Packing
Stuffing box
Stuffing
box gasket
Bonnet
Bonnet bolt
Bonnet gasket
Bonnet nut

Disc seat ring
Disc

Body seat ring

Body

NRS gate

Gate valve

Handwheel nut
Handwheel

Yoke bushigh

Stem
Gland flange
Gland
Bonnet and yoke
Packing
Stem hole bushing
Body
Disc seat ring
Locking washer

Disc
Body seat ring

Bolted bonnet, flangted

Handwheel nut
Handwheel

Stem

Gland nut
Gland

Packing

Bonnet

Bonnet nut

Plug
Body seat ring

Body

Union bonnet, thread

Globe valve

FIGURE 9.20   Globe and gate stop valves. Reproduced with permission from Crane Co.

and serve to protect equipment from over pressurization. ASME vessel code will require enough relieving capacity to ensure the vessel will not over pressurize the design pressure. Therefore all boilers will be required to have a safety relief valve(s) installed that will lift and relieve enough capacity at a specific pressure. If a boiler pressure vessel is stamped at 150 psig MAWP, any safety relief valve can be installed on that vessel provided it will relief enough energy at a pressure at or lower than 150 psig. The lifting pressure rating of the SRV is selected based on what operating range you wish to operate the steam system. For instance, a steam system the operates at 50 psig using a 150 psig boiler would have the boiler typically trimmed with a 75 psig SRV. It should be noted that steam boiler SRVs tend to start weeping at pressures up to 10% below their set point. A 75 psig SRV may leak some steam at 70 psig. The

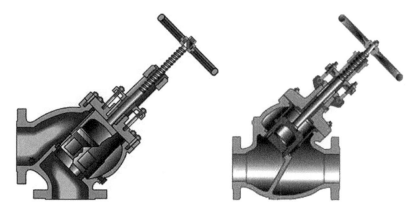

FIGURE 9.21   Straight and angle nonreturn or stop check valve. Reproduced with permission from Crane Co.

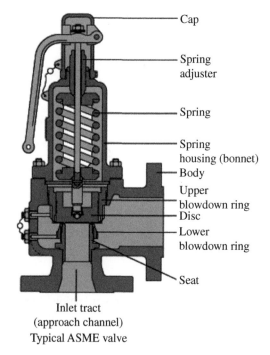

FIGURE 9.22   Safety relief valve. Spirax Sarco Inc.

SRV rating also will dictate the rest of the steam system rating. Boiler feed pumps need to be pressure rated for the boiler SRV trim pressure, not the boiler MAWP. The same holds true for downstream steam system equipment. Multiple SRVs can be used in lieu of a single one. Most large process boilers will use multiple SRVs to meet the code relieving requirements.

A typical installation of the boiler relief valves are shown in Figure 9.23. Note the ASME code requires the vessel/line opening to be equal or larger in size to the SRV inlet. You are not allowed to put an SRV with a 2 inch inlet on a vessel or line opening less than 2 inches, bushed up to 2 inches. Likewise, the code requires that the SRV discharge line cannot be bushed down to a diameter less than the SRV discharge connection size. The outlet piping of all SRVs should be vented outside whenever possible and never have a discharge point near an area that could contact vital equipment or personnel. The best design for venting an SRV is to use a drip pan elbow and vent the SRV through the roof or directly horizontally to an outside wall away from any walkways. Some jurisdictions will allow for combined SRV tail piping to be combined into one pipe; however, the common pipe must have a cross-sectional area equal or greater than the sum of all the smaller tail pipes. Designers should always check with the AHJ or Authority Having Jurisdiction for their application.

Most, if not all annual boiler inspections, require the installed SRVs on steam boilers be tested and functionality verified. Testing can be done in situ on the boiler by raising the boiler pressure and verifying the valve will lift and relieve at or below the set point rating on the valve, or the SRV can be removed and bench tested. **A good maintenance practice is to keep a spare set of SRVs for each boiler and**

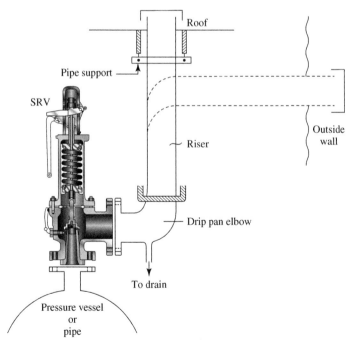

FIGURE 9.23 A typical boiler SRV installation. Data from *Crosby Pressure Relief Valve Engineering Handbook* [38].

rotate them annually and bench testing the removed prior to putting them in storage for the next rotation.

Example 9.5

A steam boiler with a steam output rating of 5600 lb/h and an MAWP of 150 psig will be operated at 75 psig. What is the correct capacity and pressure requirement of the boiler SRV.

Answer: The SRV should be sized closer to the operating pressure but, be able to relieve 5600 lb/h at the set pressure of the valve. A good selection for a SRV for this application would be a 100 psig set pressure. Consequently, it would need to be sized to relieve 5600 lb/h steam flow at 100 psig.

10

THE CONDENSATE RECOVERY SYSTEM

Steam will condense on any surface that has a lower temperature than the saturation temperature of the steam. The formed liquid from condensed steam is called **condensate.** Condensate starts to form as soon as the steam leaves the boiler. Steam headers and steam-using equipment will form condensate which will require steam traps to provide a barrier that will separate the condensate from the steam. The trapped condensate needs to be collected, discarded, or pumped back into the boiler feed water system. This part of the steam system, called the condensate recovery system, has three fundamental purposes. It must

- **separate the condensate from the steam,**
- **collect the condensate, and**
- **pump the condensate into the feed water tank or be discharged.**

The condensate formed in both the steam distribution pipe work and in the process equipment is a convenient supply of useable hot boiler feed water. Although it is important to remove this condensate, it is a valuable commodity and should not be allowed to run to waste. Returning all condensate to the boiler feed tank closes the basic steam loop and should be practiced wherever practical.

A general piping and instrumentation diagram of a typical condensate is shown in Figure 10.1. Condensate recovery systems can be divided into the five subsystems listed below, with each subsystem requiring its own design considerations. Not all condensate recovery systems use all five subsystems, but all will have drain lines and steam traps.

Process Steam Systems: A Practical Guide for Operators, Maintainers, Designers, and Educators, Second Edition. Carey Merritt.
© 2023 John Wiley & Sons, Inc. Published 2023 by John Wiley & Sons, Inc.

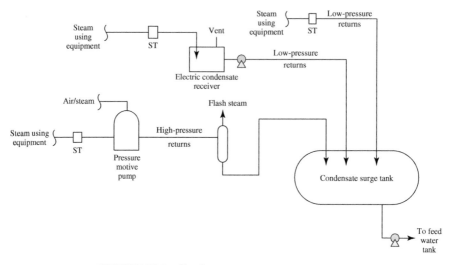

FIGURE 10.1   Condensate recovery system concept.

The five subsystem are

a) Drain lines from the steam-using equipment to the traps carry the condensate, pressurized high-temperature hot water.
b) Steam traps and trap discharge lines carry a two-phase mixture of flash steam and condensate.
c) Flash tanks that reduce the hot pressurized water to atmospheric or near atmospheric pressure.
d) Condensate collection and pumped return systems utilizing electric or nonelectric pumps.
e) Surge tank to accept the condensate from different sources and provide a means to store water for the feed water tank.

This hot condensate is essential for energy recovery and is demonstrated in Example 10.1 below.

Example 10.1
A steam boiler produces 20,000 lb/h at 120 psig steam and none of the condensate is returned back to the boiler. Consequently, the boiler uses 100% fresh makeup water at 70 F. A plant modification can be done for $100,000 that would allow 70% of the condensate to be returned to the boiler. Is the modification worth doing?

Answer: To determine the viability of this modification, you must determine the cost savings from returning the condensate and weigh that against the cost of the modification. You will also need to know the temperature of the condensate (if unknown, then assume it will be 200 F) and the cost of fuel (we assume natural gas *at $5/million btu). First, determine the amount of heat energy required to heat the boiler water with and without the condensate. We know that 120 psig water will boil at 350 F.*

Without condensate: 20,000 lb/h × 1 btu/lb-F × (350 F – 70 F) = 5,600,000 btu/h. With 70% condensate the new boiler water temperature can be calculated by adding the portions of each of the 70 F makeup water and the 200 F condensate. Therefore, the new boiler water temperature would be (70 F × 30%) + (200 F × 70%) = 161 F. So with 70% condensate returned, the heat energy needed to raise the boiler water from 161 to 350 F will be, 20,000 lb/h × 1 btu/lb-F × (350 F – 161 F) = 3,780,000 btu/h. Then the energy savings per hour is 1,820,000 btu/h, and at $5/million btu, we can save 1.82 × 5 = $9.10/h. This doesn't seem like a lot, but annualized at 24/7 the saving is $78,840/year. Furthermore, if we take into the efficiency of the boiler (around 80%) then this savings become even greater (i.e., $78,840/0.80 = $98,550/ year). The actual savings would depend on how many equivalent rated hours of operation the boiler achieved in a year. When you consider that a steam system is designed to operate for 20 years, the modification with a one year payback is well worth doing.

CONDENSATE LINE SIZING

Drain lines to traps. The condensate has to flow from the condensing surface to a steam trap. In most cases, this means that gravity helps to induce flow, since the heat exchanger steam space and the traps are at the same pressure. The lines between the drainage points and the traps should have a minimum slope of 1″ in 10 feet toward the trap. Table 10.1 shows the water carrying capacities of the pipes with such a gradient. It is important to allow for the passage of noncondensable gases to the trap, and for the extra water to be carried at cold starts. **In most cases, it is sufficient to size the drain pipes to handle 1.5 to 2 times the condensate produced at full running load**. Unless, drain lines from the steam-using equipment to the trap are long (i.e., >50 ft), friction losses should be minimal.

TABLE 10.1 Condensate Flow Versus Pipe Size in lb/h

Steel	Approximate Frictional Resistance			
Pipe	In Inches Water Column per 100 ft of Travel			
Size	1	5	7	10
½″	100	240	290	350
¾″	230	560	680	820
1″	440	1070	1200	1550
1 ¼″	950	2300	2700	3300
1 ½″	1400	3500	4200	5000
2″	2800	6800	8100	9900
2 ½″	5700	13,800	16,500	20,000
3″	9000	21,500	25,800	31,000
4″	18,600	44,000	52,000	63,400

Data from Spirax Sarco [16].

Trap discharge lines. At the outlet of steam traps, the condensate return lines must carry condensate, noncondensable gases and flash steam released from the condensate. Where possible, these lines should drain by gravity to the condensate receiver, whether this be a flash recovery vessel or the vented receiver. When sizing return lines, two important practical points must be considered. First, one pound of steam has a specific volume of 26.8 cubic feet at atmospheric pressure. It also contains 970 btu of latent heat energy. This means that if a trap discharges 100 pounds per hour of condensate from 100 psig to atmosphere, the weight of flash steam released will be 13.3 pounds per hour, having a total volume of 356.4 cubic feet. It will also have 12,901 btu of latent heat energy. This will appear to be a very large quantity of steam and may well lead to the erroneous conclusion that the trap is passing live steam (failed open).

Secondly, the actual formation of flash steam starts to take place downstream of the steam trap orifice where pressure drop occurs. From this point onward, the condensate return system must be capable of carrying some flash steam, as well as condensate. **Sizing of condensate return lines from trap discharges based totally on water is a gross error and causes lines to be undersized for the flash steam**. This causes condensate lines to become pressurized, not atmospheric, which in turn causes a backpressure to be applied to the trap's discharge which can cause equipment failure and water logging. The sizing chart shown in Figure 10.2 can be used to size condensate lines.

When flash steam volume is not accounted for, a positive pressure can develop in the condensate return system by the flash steam. The condensate return line will follow the pressure/temperature relationship of saturated steam. So, trap testing showing elevated downstream condensate return temperatures does not necessarily mean a trap has failed. It may be due to undersized condensate lines. When sizing condensate return lines, the volume of the flash steam must be considered. The chart below allows the lines to be sized considering flash steam. By determining the quantity of flash steam and sizing the return line for velocities between 4000 and 6000 ft/min, the two-phase flow within the pipe can be accommodated.

Draining condensate from traps serving loads at differing pressures to a common condensate return line can be accomplished. At the downstream or outlet side of the traps, the pressure must be at common pressure in the return line. This return line pressure will be the sum of at least three components.

1. The pressure at the end of the return line, either atmospheric or of the vessel into which the line discharges.
2. The hydrostatic head needed to lift the condensate up any risers in the line.
3. The pressure drop needed to carry the condensate and any flash steam along the line.

Item 3 is the only one likely to give rise to any problems if condensate from sources at different pressures enters a common line. The return line should be sufficiently large to carry all the liquid condensate and the varying amounts of flash steam associated with it, without requiring excessive line velocity and excessive pressure drop.

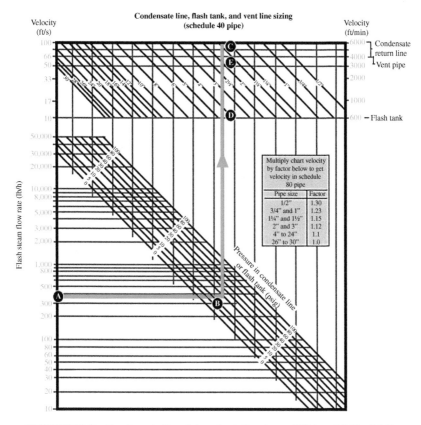

FIGURE 10.2 Condensate line sizing chart. Courtesy of Watson McDaniel Co.

Example 10.2

Size a condensate return line from a 125 psig steam system trap discharging 3000 l/h hot condensate to 5 psig flash tank.

*Answer: Determine percent flash steam produced using Table 4.1, Flash Steam Chart A steam pressure of 125 psig and a flash tank pressure of 5 psig will result in 13.5% of the condensate flashing to steam. Next, multiply the condensate load by the percent flash from step #1 to determine the flow rate of flash steam produced. 3000 lb/h × 0.135 = 405 lb/h. Now from Figure 10.2 with the flash steam flow rate of 405 lb/h at "A" and move horizontally to the right to the flash tank pressure of 5 psig "B." Rise vertically to choose a condensate return line size which will give a velocity between 4000 and 6000 ft/min, "C." In this example, a **2 inch schedule 40 pipe** will have a velocity of approximately 5000 ft/min. If schedule 80 pipe is to be used, refer to table within body of chart. Multiply the velocity by the factor to determine whether the velocity is within acceptable limits.*

STEAM TRAP APPLICATIONS

Steam traps come in a variety of sizes and functionality. Applying the correct size, pressure rating, and type of trap is essential to a good condensate recovery system design. Incorrect application can lead to system heat transfer inefficiencies and wasted money. Likewise, trap location relative to the heat exchange equipment is essential to ensuring proper condensate draining. This discussion of steam traps reviews the type of steam traps and their proper usage. There is no such thing as a "universal" steam trap which is suitable for all applications. For this reason, you should familiarize yourself with each of the main steam trap groups and learn how best take advantage of the merits of each type.

When the steam system is shut down, air will be drawn in to take up the space formerly occupied by steam. Since this air has to be removed from the system on startup, it is a considerable bonus if the steam traps have a good air venting capability. While this is the case with certain traps, other types are actually prone to "air binding" – a condition in which the trap remains closed when it should be opening to release condensate. There are three main steam trap groups:

Thermostatic Group

This group separates steam and condensate by the temperature difference of each which operates a thermostatic, valve-carrying element. Condensate must cool below steam temperature before it can be released.

Mechanical Group

Traps of this group operate an internal mechanical device, sensing the difference in density between steam and condensate. The movement of a "float" or a "bucket" operates a valve to discharge condensate.

Thermodynamic Group

This group works on the difference in kinetic energy or velocity between steam and condensate flowing through the trap. The most widely used types are thermodynamic disc models. In these traps, the valve consists of a simple disc which closes to high velocity steam but opens to lower velocity condensate.

Thermostatic Steam Trap Group

Balanced pressure type. A typical balanced pressure thermostatic steam trap is shown in Figure 10.3. The thermostatic element will expand and contract if pulled or pushed at the sealed ends. The element is filled with a liquid (alcohol mixture) which has a boiling point lower than that of water. Air and cooler condensate will be pushed out through the wide open valve. As the condensate gradually warms up, heat transfer will take place to the alcohol mixture inside the element. Before the condensate

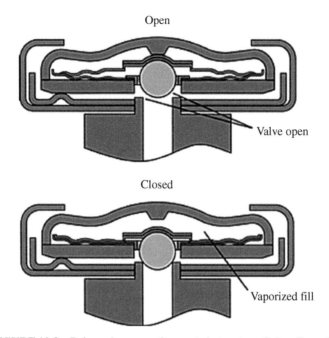

FIGURE 10.3   Balanced pressure thermostatic type trap. Spirax Sarco Inc.

reaches steam temperature, the mixture reaches its boiling point. As soon as it boils, vapor is given off, increasing the pressure inside the element. The pressure inside the element exceeds the pressure in the trap body and the element expands forcing the trap to close off flow. Eventually the condensate in the trap body cools down, allowing element to contract and open the valve to condensate flow. Condensate is then discharged through the open valve and the complete cycle is repeated. Thermostatic balanced pressure traps are small, light, and have a large capacity for their size. This type of trap is unlikely to freeze when working in an exposed position. The balanced pressure trap automatically adjusts itself to variations of steam pressure up to the maximum pressure for which it is suitable. Trap maintenance is easy. The element and valve seat are detachable and can be replaced in a few minutes without removing the trap body from the line. The flexible element in this type of trap may be susceptible to damage by water hammer or corrosive condensate.

Liquid expansion thermostatic steam trap. A typical liquid expansion trap, as shown in figure 10.4, is operated by the expansion and contraction of a liquid-filled well which responds to the temperature difference between steam and condensate. As the temperature of the condensate passing through the trap increases, heat is transmitted to the well causing it to expand. This expansion acts on the piston and the valve is pushed nearer and nearer to its seat, steadily reducing the flow of condensate. When steam reached the well, the valve is shut completely off. If the cooler condensate is formed, the valve will reopen and pass just this amount. This type of trap can be adjusted to discharge at very low temperatures, if so desired. Like the balanced

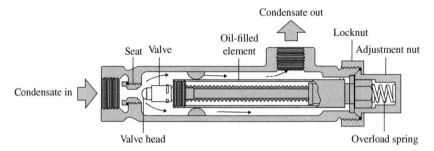

FIGURE 10.4   Liquid expansion thermostatic-type trap. Spirax Sarco Inc.

pressure trap, the liquid expansion trap is fully open when cold giving good air discharge and maximum condensate capacity on startup loads. If the steam pressure at the trap is subject to wide and rapid variation, the element will not respond to the changes as quickly as that of a balanced pressure trap. The flexible tubing of the element can be destroyed by corrosive condensate. Since the liquid expansion trap discharges condensate at a temperature of 212 F or below, it should never be used on applications which demand immediate removal of condensate from the steam space.

Bimetal-type steam trap. Condensate flow in this type of steam trap results from the bending of a composite strip of two metals which expand by a different amount when heated up. One end of the strip is fixed to the trap body, while the other end is connected to the valve. Air and condensate pass freely through the open valve until the bimetal strip approaches steam temperature. The free end will then bend downward and close the valve. The trap will remain shut until the body is filled with condensate which has cooled sufficiently to allow the strip to straighten and open the valve. Bimetallic traps are usually small in size and yet can have a large condensate discharge capacity. The valve is wide open when the trap is cold, giving a good air venting capability and maximum condensate discharge capacity under startup conditions. Bimetallic traps, like the one shown in Figure 10.5, can be constructed to withstand water hammer, corrosive condensate, high steam pressures, and superheated steam. Bimetallic traps do not usually respond quickly to changes in load or pressure because the bimetal is relatively slow to react to variations in temperature. As condensate is discharged below steam temperature, waterlogging of the steam space will occur unless the trap is fitted to the end of a fairly long cooling leg.

Mechanical Group

Float and lever type. With a simple float and lever-type steam trap, the condensate enters the trap body through the inlet and the ball lifts as the water level rises. The float arm connects the ball to the outlet valve which is gradually opened as condensate raises the ball. The position of the valve varies according to the level of water in the trap body, giving continuous condensate discharge on any load which falls within the maximum capacity of the trap. If the condensate load diminishes and steam

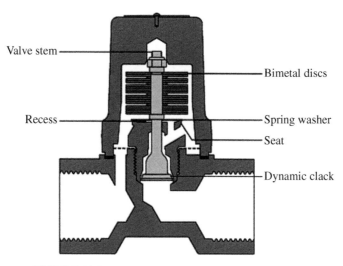

FIGURE 10.5   Bimetal-type steam trap. Spirax Sarco Inc.

reaches the trap, the float will drop to its lowest position. The valve is held firmly against its seat and no steam can be wasted. The one major drawback with this trap is that air cannot be discharged through the main valve on startup. Unless some means is provided for releasing air from the system, condensate will be prevented from flowing into the trap, which then becomes "air-bound." A design feature that allows for automatic air venting is the addition of a thermostatic element. This is, in fact, a thermostatic element of the type used in the thermostatic traps already explained in this section. These types of traps are called **float and thermostatic types**. The thermostatic valve is wide open when the trap is cold, so that the air is readily discharged on startup. As soon as steam reaches the trap, the element expands and pushes the valve shut so no steam is able to escape. The float and lever trap gives continuous discharge of condensate at steam temperature. This makes it the first choice for applications where the rate of heat transfer is high for the area of heating surface available. It is able to handle heavy or light condensate loads equally well, and it is not adversely affected by wide and sudden fluctuations of pressure. Floats, bellows, and the thermostatic elements, however, are susceptible to damage by water hammer. The materials of this type of thermostatic element cannot tolerate corrosive condensate and they are not suitable for use on superheated steam, unless modified. A float trap can be damaged by freezing and the body should be well insulated if it is to be placed in an outdoor location where freezing conditions exist. A typical float- and thermostatic- type trap is shown in Figure 10.6.

Open top bucket type. An open top bucket can be used to operate the valve instead of a ball float. The bucket will float in condensate when empty but sink by its own weight when full of condensate. When condensate enters, it first fills the body of the trap outside the bucket. The bucket floats and the valve is pushed up on to its seat. More condensate flowing into the body spills over into the bucket. When it is full

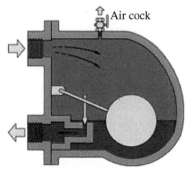

Float trap with air cock

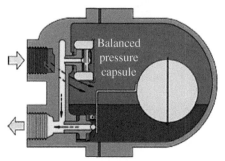

Float trap with thermostatic air vent

FIGURE 10.6   Float-type trap with manual and thermostatic-type air venting. Spirax Sarco Inc.

enough, the bucket is sufficiently heavy to drop back to the bottom of the trap, drawing the valve away from its seat. The steam pressure on the condensate in the trap forces water out through the central tube and the bucket becomes buoyant once again. The whole action is then repeated. It should be noted from this description that traps of this type have an intermittent blast discharge action. Open bucket traps are usually robust and can be made for use on high pressure and superheated steam. They can withstand water hammer and corrosive condensate better than most types of mechanical traps and there is little that can go wrong with the simple mechanism. This mechanical limitation means that open bucket traps tend to be rather large and heavy in relation to their discharge capacity. No provision is made for air venting unless either a manual cock or thermostatic air vent is fitted. This type of trap is susceptible to damage by freezing and the body must be well insulated if it is placed outdoors.

Inverted bucket type. A trap which is more commonly used than the open bucket is the inverted bucket pattern. A typical inverted bucket type trap is shown in Figure 10.7. In this type, the operating force is provided by steam entering an inverted bucket and causing it to float in the condensate with which the trap is filled. When steam is turned on, the bucket is at the bottom of the trap and the valve is wide open. Air is discharged through a small hole in the top of the bucket. Condensate enters the trap and the water level rises both inside and outside the bucket. The bucket remains at the bottom of the trap and the water is able to pass away through the wide open valve. When

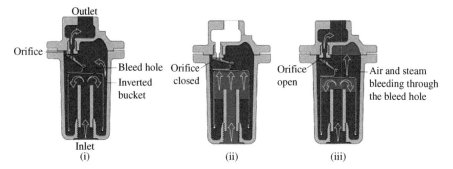

Outlet

Orifice

Bleed hole
Inverted
bucket

Orifice
closed

Orifice
open

Air and steam
bleeding through
the bleed hole

Inlet
(i)

(ii)

(iii)

FIGURE 10.7 Inverted bucket trap operation. Spirax Sarco Inc.

steam reaches the trap, it enters the bucket and makes it float upward, shutting the valve through a lever arrangement. The steam in the bucket will slowly escape via the small vent hole, collecting at the top of the trap. If it is replaced by more steam, the trap remains closed. If condensate enters, the bucket will sink, pulling the valve open. Like the open bucket trap, this type has an intermittent blast discharge action. The inverted bucket trap can be made to withstand high pressures and can be used on superheated steam if a check valve is fitted on the inlet. It has a reasonably high degree of tolerance to water hammer conditions and there is little to go wrong with the simple bucket and lever mechanism. There should always be enough water in the trap body to act as a seal around the lip of the bucket. If the trap loses this water seal, steam can blow to waste through the outlet valve. If an inverted bucket trap is used on an application where fluctuation of the pressure can be expected, a "check valve" or a "nonreturn valve" should be fitted on the inlet line in front of the trap. The inverted bucket trap is likely to suffer damage from freezing if placed outdoors in an exposed position.

Thermodynamic Group

Thermodynamic disc type. The construction of the thermodynamic disc type of steam trap is extremely simple. A typical trap, as shown in Figure 10.8, consists of only a body, a top cap, and a free floating disc. This disc is the only moving part of the trap. An annular groove is machined into the top of the trap body, which forms the seat face. The faces of the seat and of the disc are carefully ground flat, so that the disc seats on both rings at the same time. This seals off the inlet from the outlet and is essential if a tight shut off is to be achieved. On startup, air and cool condensate reach the trap, passing up the inlet orifice. Air and condensate flow radially outward from the center of the disc into the space between the seat rings and are discharged through the outlet passage. The temperature of the condensate gradually increases and as this passes through the trap inlet, some of it flashes into steam. The resulting mixture of flash steam and condensate flows radially outward across the underside of the disc and because flash steam has a larger volume than the same weight of condensate, the speed of flow increases steadily as more and more flash steam is formed. In order to understand what happens next, it is necessary to have a basic grasp of what is known as "Bernoulli's Theorem." This simply states that in a moving fluid, the

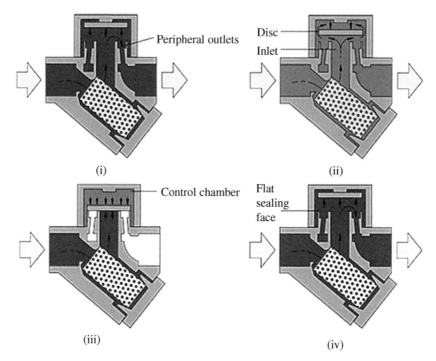

FIGURE 10.8   Thermodynamic-type trap shown with integral strainer. Spirax Sarco Inc.

total pressure is the same at all points. This total pressure is the sum of the static and dynamic pressures of a fluid. The static pressure is that which would be measured by a pressure gauge, while the dynamic pressure is that which would be produced by the individual fluid particles if they were to be brought to rest by hitting an obstruction. The dynamic pressure increases as the speed of the particles increases.

If we apply this theorem to the thermodynamic disc trap, we can appreciate that the dynamic pressure of the steam and condensate flowing under the disc will increase as the speed of flow increases. Since the total pressure must remain constant, the static pressure falls as the dynamic pressure rises. The flash steam exerts a static pressure on the whole of the top surface of the disc. This builds up until it is sufficient to overcome the inlet pressure which acts only on a small section in the middle of the disc. When this happens, the disc snaps shut against the seat rings preventing further flow through the trap. The disc remains firmly against its seat until the flash steam above it condenses. This relieves the pressure acting on the top of the disc, allowing it to be raised again by the inlet pressure. If there is no condensate waiting to be discharged when the trap opens, a small amount of high pressure steam will enter the control chamber and cause the disc to seat very quickly. The addition of a strainer helps to prevent the possibility of dirt particles either blocking the small outlet holes of the trap or preventing the disc from giving a tight shut off. Thermodynamic disc traps can operate within their whole working range without any adjustment or change of valve size. They are compact, simple, lightweight, and have a large condensate handling capacity for their size. This type of trap can be used on high pressure and superheated steam and is not damaged by water hammer or vibration. They are usually stainless steel construction that offers a high degree of resistance to corrosive condensate. As the disc is the only moving part, maintenance can easily be carried out without removing the trap from the line. The disc prevents any return flow of condensate through the trap, cutting out the need for a separate check valve. When failed open, the disc moves rapidly up and down giving off a "clicking" sound. This sound is an audible warning that tells us trap maintenance is required. Thermodynamic traps will not work positively on very low inlet pressures or high back pressures.

Sizing a steam trap can also be done by applying some basic information [2]. The flow chart shown in Figure 10.9 can be used as a guide.

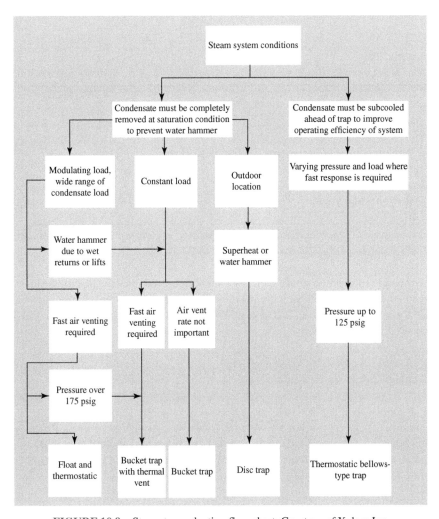

FIGURE 10.9   Steam trap selection flow chart. Courtesy of Xylem Inc.

Required information:

1. The steam pressure at the trap: After any pressure drop through control valves or equipment.
2. The lift, if any, after the trap. Rule of thumb: 2.3 ft head. = 1 psi back pressure.
3. Any other possible sources of back pressure in the condensate return system. For example, condensate taken to a pressurized DA tank or back pressure due to discharges of numerous traps close together into small sized return.
4. Quantity of condensate to be handled. Obtained from calculation of heat load.
5. *Safety factor*: These factors depend upon particular applications. **Rule of thumb**: Use factor of 2 on everything except temperature-controlled air heater coils and converters, and siphon applications.

Example 10.3
A trap is required to drain 45 lb/h of condensate from a 6" insulated steam main, which is supplying steam at 100 psig. There will be a lift after the trap of 20 ft.

Supply pressure = 100 psig
Lift = 20 ft = 9 psig
Therefore, differential pressure = 100 – 9 = 91 psig
Quantity = 45 lb/h
Mains drainage factor = 2
Therefore, sizing load = 90 lb/h

Answer: A small reduced capacity thermodynamic steam trap will easily handle the 90 lb/h sizing load at a differential pressure of 90 psi.

If maximum steam system efficiency is to be achieved, the best type of steam trap must be fitted in the most suitable position for the application in question, the flash steam should be utilized, and the maximum amount of condensate should be recovered.

FLASH STEAM UTILIZATION

We discussed in Chapter 4 that hot pressurized water can flash to steam when it is depressurized. Likewise, hot pressurized condensate discharged from a steam trap will create flash steam. If the flash steam is to be recovered and utilized, it has to be separated from the condensate. This is best achieved by passing the mixture of flash steam and condensate through what is known as a "flash tank" or "flash vessel."

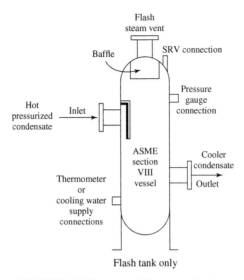

FIGURE 10.10   Typical flash tank design.

A typical flash tank is shown in Figure 10.10. These vessels are usually ASME Section VIII pressure vessels rated for 50 psig or less. When designing a steam system with multiple steam pressure users, dumping all condensate into a flash tank will eliminate any consequences of trying to transport various pressure condensate. The flash tank is normally located between the steam traps and condensate collection system. Figure 10.10 shows a typical flash tank design

The size of the vessel has to be designed to allow for reduced steam velocity so that the separation of the flash steam and condensate can be accomplished adequately. It must also prevent carryover of condensate into the flash steam recovery system. The target tank steam velocity is 600 ft/min to ensure proper separation. The condensate drops to the bottom of the flash tank where it is discharged directly into the condensate collection system. The flash steam outlet connection is sized so that the flash steam velocity through the outlet is approximately 3600 ft/min. The condensate inlet should also be sized for <6000 ft/min flash velocity.

How to Size Flash Tanks and Vent Lines

Whether a flash tank is atmospheric or pressurized for flash recovery, the procedure for determining its size is the same. The most important dimension is the diameter. It must be large enough to provide adequate separation of the flash and condensate to minimize condensate carryover. Figure 10.11 can be used to size the tank and openings.

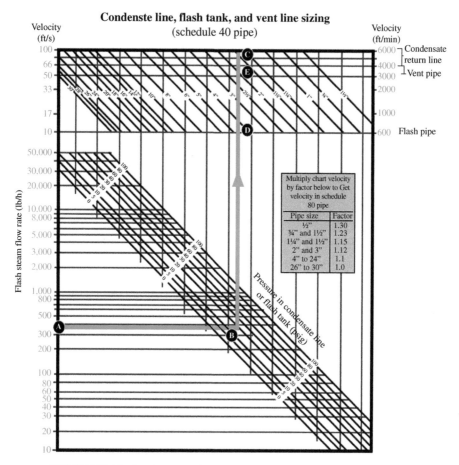

FIGURE 10.11 Flash tank design graph. Courtesy of Watson McDaniel Co.

Example 10.4

Size a flash recovery vessel receiving condensate from a 125 psig steam trap discharging 4000 lb/h into the condensate return system at 0 psig.

Answer:

1. *Determine percent flash steam produced using Figure 2.6. With a steam pressure of 125 psig and a flash tank pressure of 0 psig, 14.9% of the condensate will flash off.*
2. *Next, multiply the condensate load by the percent flash from Step #1 to determine the flow rate, of flash steam produced: 4000 lb/h × 0.149 lb/h = 596 lb/h.*
3. *Using the calculated flash steam quantity of 596 lb/h enter Figure 9.11 at "A" and move horizontally to the right to the flash tank pressure of 0 psig "B." Rise vertically to the flash tank diameter line (600 ft/min) at "D." The tank diameter of 24" is required.*

4. *From point "D" continue to rise vertically to "E" to determine the size of vent pipe to give a velocity between 3000 and 4000 ft/min. In this case 10" schedule 40 pipe connection is required.*

5. *The condensate inlet connection size can also be determined using the same graph. As in step 4 find the vertical line that you determined the vent connection size and continue up the graph until the vertical line intersects the horizontal 100 ft/s velocity line (actually 6000 ft/min). The graph shows a 8" condensate line connection is adequate.*

The answer is a vessel that has a diameter of 24" and has a 10" vent with a 8" condensate inlet connection. Generally, a flash tank will have a 2:1 aspect ratio height to width. Therefore, this vessel would likely be a 48" × 24" vessel.

In an efficient and economical steam system, this so-called flash steam will be utilized on any load which will make use of low-pressure steam. Sometimes it can be simply piped into a low-pressure distribution main for general use. Flash steam can be used whereever low-pressure steam is needed. Preheating process fluid, pressurizing a feed water deaerator, or using the steam to make hot water are some typical applications for flash steam utilization. The flow diagram below in Figure 10.12 shows flash steam usage for a variety of applications.

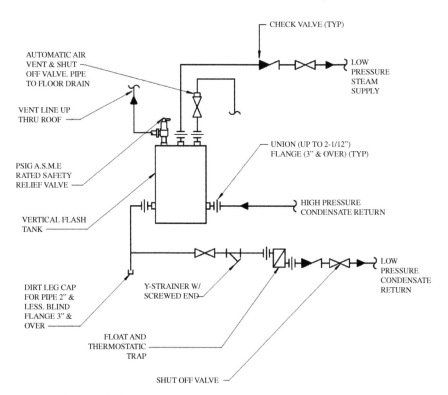

FIGURE 10.12   Flash steam recovery detail. Courtesy of C&S Engineers.

Flash steam recovery is simplest when being recovered from a single piece of equipment that condenses a large amount of steam, such as a large steam kettle or process heat exchanger. The flash steam recovery system by design will apply a backpressure to the equipment being utilized as the flash steam source. Only as a last resort should flash steam be vented to atmosphere and lost.

CONDENSATE COLLECTION

The trapped condensate must be collected and pumped back to the feed water system if it is to be reused. Condensate collection will vary in design and is dependent on the size of the steam system. Smaller steam plants (i.e., 100 hp and smaller) may have the condensate flow directly back to the feed water tank. In these applications it is recommended the feed water tank be oversized at least 50% to account for slugs of condensate returning which may create a water inventory problem on startup. Remember, as a steam system is allowed to cool down the steam will condense in the steam piping and sit there until the system is either drained manually or repressurized. Large low-pressure steam systems will often see significant amounts of condensate slug back to the feed water or surge tank during start up after a system shutdown.

Steam systems having a large distribution network can utilize a series of condensate collectors and a surge tank. The condensate collectors or drainers as they are sometimes called, are usually located near the steam traps. They are usually small (i.e., 15 gallon or less) and will cascade to the condensate surge tank. The surge tank is used as the common collection tank and typically fed directly to the boiler feed tank or deaerator.

Electric Condensate Return System

When using electric pumps to lift the condensate, packaged units comprising of a receiver tank (usually vented to atmosphere) and one or more motorized pumps are commonly used. It is important with these units to make sure that the maximum condensate temperature specified by the manufacturer is not exceeded, and the pump has sufficient capacity to handle the load. Condensate temperature usually presents no problem with returns from low-pressure steam systems. There, the condensate is often below 212 F and a little further subcooling in the gravity return lines and the pump receiver itself provides little difficulty in meeting the maximum temperature limitation.

On high-pressure systems, the gravity return lines often contain condensate at just above 212 F, together with some flash steam. If a flash tank is not used, then the condensate must be cooled to <200 F or pump cavitation will likely occur. Uninsulated condensate piping to the collector can provide some condensate cooling. The condensate water must remain in the receiver for an appreciable time if it is to cool sufficiently, or the pump discharge may have to be throttled down to reduce the pump's capacity to avoid cavitation. In some cases the pumps are supplied coupled to receivers and the static head above the pump inlet is already fixed by the pump manufacturer; it is only necessary to ensure that the pump set has sufficient capacity at the water temperature expected at the pump. Pump manufacturers usually have a set of capacity curves for the pump when handling water at different temperatures and these should be consulted. Condensate receivers that receive hot pressurized condensate and have to pump the condensate with low lift requirements will be highly susceptible to pump cavitation. The receiver must be vented or a flash tank used in front of the receiver to reduce the condensate to atmospheric pressure. Typical condensate receivers are shown below in Figures 10.13 and 10.14.

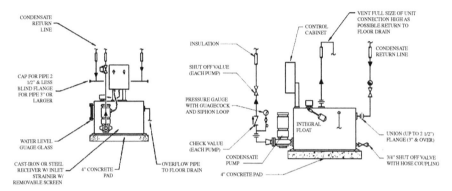

FIGURE 10.13 Electric condensate return system piping detail. Courtesy of C&S Engineers.

FIGURE 10.14 Larger condensate tank. Shippensburg Pump Co.

Pressure Motive Condensate Pump

Condensate can also be moved by the use of nonelectric condensate pumps, such as the pressure motive pump. The pressure motive pump, sometimes called a **pump trap**, is essentially an alternating receiver which can be pressurized, using steam, air, or other gas. The gas pressure displaces the condensate (which can be at any temperature up to and including boiling point). Check valves at the inlet and outlet of the pump body ensure condensate can enter and exit the receiver at the right time.

When the receiver is full of condensate, an internal float mechanism opens the steam or pressurized supply gas valve. This pressurizes the receiver and forces the inlet check valve to close and opens an exhaust check valve. The pressurizing gas forces (i.e., pumps) the condensate out of the receiver. When the receiver is emptied, the float falls and closes the pressurized gas supply allowing the discharge check valve to close and the inlet check valve to reopen. The pressurized gas is vented to atmosphere or to the space from which the condensate is being drained as the receiver fills with condensate. When the pressures are equalized, condensate can flow by gravity into the pump body to refill it and complete the cycle. As the pump fills by gravity only, there can be no cavitation and this pump readily handles boiling water or other liquids compatible with its materials of construction. A summary of its operation is shown in Figure 10.15.

The capacity of the pump depends on the filling head available, the size of the condensate connections, the pressure of the operating steam or gas, and the total head through which the condensate is lifted. This will include the difference in elevation between the pump and the final discharge point, any pressure difference between the pump receiver and final receiver; friction in the connecting pipe work, and the force necessary to accelerate the condensate from rest in the pump body up to velocity in the discharge pipe. Tables listing capacities under varying conditions are usually provided in the equipment manufactures catalog bulletins.

Pressure Motive Pump Installation Requirements

Depending upon the application, the pressure motive pump body is piped so that it is vented to atmosphere or, in a closed system, is pressure equalized back to the space that it drains. This allows condensate to enter the pump except during the short discharge stroke when the inlet check valve is closed and condensate accumulates in the inlet piping.

To eliminate the possibility of condensate backing up into the steam space, reservoir piping must be provided above the pump with volume as specified by the manufacture. A closed system requires only a liquid reservoir. In open systems, the vented receiver serves this purpose as it is always larger in order to also separate the flash steam released. A typical installation of an open or closed loop PMP system is shown in figure 10.16.

Vented or open systems. Condensate from low-pressure steam systems may be piped directly to a small size pressure motive pump only when 50 lb/h or less of flash steam must vent through the pump body. This does not eliminate the requirement that there must be enough piping to store condensate during the brief discharge cycle. In many low-pressure systems, the reservoir may be a section of large horizontal pipe which is vented to eliminate flash steam. In higher pressure, high load systems, the larger quantity of flash released requires a vented receiver with piping adequate to

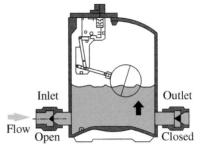

① Condensate flows from the receiver tank through the inlet check valve and fills the pump tank. During the filling cycle, the float inside the tank rises

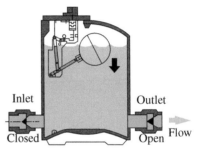

② When the pump tank has filled to the trip point, the mechanism triggers, opening the motive gas inlet valve and simultaneously closing the vent valve. This allows motive pressure to enter the pump body, which drives the condensate through the outlet check valve into the condensate return line. During the discharge cycle, the liquid level and the float inside the pump tank drop

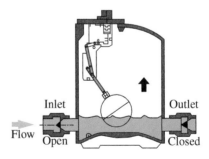

③ At the lower trip point, the mechanism triggers and the motive gas inlet valve to the pump tank closes and simultaneously the vent valve opens. The fill and discharge cycle then repeats itself

FIGURE 10.15 Steam pressure motive pump operation. Watson McDaniel Co.

permit complete separation. To prevent carryover of condensate from the vent line, the receiver should be sized to reduce flow velocity to about 600 ft/min.

Closed loop systems. It is often advisable where larger condensate loads are being handled to dedicate a pressure motive pump to drain a single piece of equipment. The

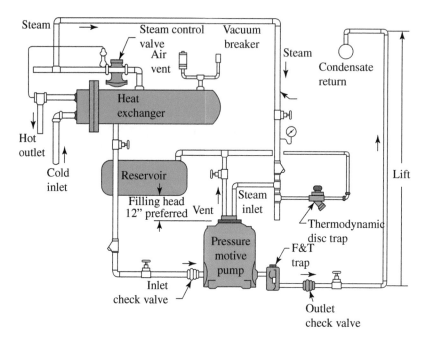

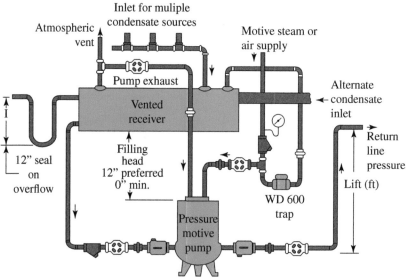

FIGURE 10.16 Installation of PMP in open and closed loop systems. Watson McDaniel Co.

pump exhaust line can then be directly connected to the steam space of a heat exchanger or, preferably with air heating coils, to the reservoir. This allows condensate to drain freely to the pump inlet and through a steam trap at the pump outlet. Only liquid is contained in the reservoir of a closed loop system. The pressure-motive pump functions as a pumping trap, and the steam supply must be greater than the return line pressure.

Pumped Condensate Return Line Installation

Finally, the condensate is often pumped from the receiver(s) to the boiler plant. These pumped condensate lines carry only water, and high water velocities can often be used so as to minimize pipe sizes. The extra friction losses entailed must not increase back pressures to the point where the pump capacity is affected. Velocities in pumped returns should be limited to 6–8 ft/s. Electric pumps are commonly installed with pumping capability of 2-1/2 or 3 times the rate at which condensate reaches the receiver. This increased instantaneous flow rate must be kept in mind when sizing the delivery lines. Similar considerations apply when steam powered pumps are used, or appropriate steps taken to help attain constant flow along as much as possible of the system.

Where long delivery lines are used, the water flowing along the pipe as the pump discharges attains a considerable momentum. As the end of the discharge cycle when the pump stops, the water tends to keep moving along the pipe and may pull air or steam into the delivery pipe through the pump outlet check valve. When this bubble of steam reaches a cooler zone and condenses, the water in the pipe is pulled back toward the pump. As the reversed flow reaches and closes the check valve, water hammer often results. This problem is greatly reduced by adding a second check valve in the delivery line some 15 or 20 ft from the pump. If the line lifts to a high level as soon as it leaves the pump, then adding a vacuum breaker at the top of the riser is often an extra help. However, it may be necessary to provide means of venting from the pipe at appropriate points, the air which enters through the vacuum breaker.

The practice of connecting additional high-pressure steam trap discharge lines into the pumped main is to be avoided whenever possible. **The flash steam which is released from this hot condensate leads to a thermal shock wave creating a banging noise within the piping commonly associated with steam hammer**. The traps should discharge into a separate gravity line which carries the condensate to a vented receiver. If this is impossible, an alternative method is to pipe the trap discharge through a sparge or diffuser inside the pumped return line. A typical diffuser installed in a return line is shown in Figure 10.17. The trap most suitable for this application would be the

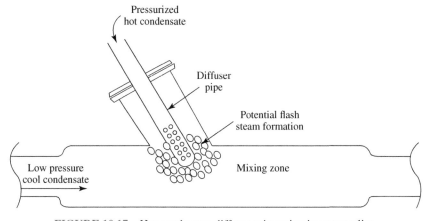

FIGURE 10.17   Hot condensate diffuser orientation in a return line.

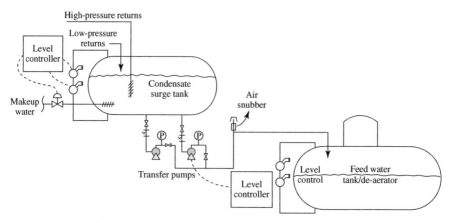

FIGURE 10.18   Surge tank piping integration with the feed water tank.

float and thermostatic type due to its continuous discharge. This is very much a compromise and will not always avoid the noise although it should reduce the severity.

SURGE TANK APPLICATION

In nearly all steam-using plants, condensate must be pumped from the location where it is formed back to the boiler house and be lifted into a boiler feed tank or deaerator. If multiple condensate collectors are used, it is often desirable to cascade the combined pump discharges to a surge tank. A surge tank is a tank located between the boiler house feed water deaerator and condensate collectors. Its main function is to act as a means to level out the flow of condensate and fresh makeup water to the feed water system. Typically a surge tank is about half the size of the feed water tank and will have transfer pumps that come on and off based on the feed water tank level. Surge tanks can be ASME code or noncode tanks but generally are made of stainless steel to handle the low pH condensate water. If the surge tank is located near several steam using pieces of equipment, the trapped condensate lines can cascade directly to a flash tank bolted directly on the top of the surge tank and let the low-pressure condensate gravity drain directly into the surge tank. Similarly when the surge tank is located relatively close to the feed water tank, then the flash steam can be used to preheat fresh make up water or pressurize the feed water DA. A flow diagram is shown below in Figure 10.18. Since the surge tank is a condensate collector, the only level control is to allow fresh make up water in on a low-level signal.

11

THE FEED WATER SYSTEM

A boiler feed water system receives the condensate from the condensate system and stores the condensate in a feed water tank. Once the condensate reaches the feed water tank it becomes feed water. The feed water system prepares the water for reintroduction into the boiler. This preparation includes, collection, deaeration, and pumping the water to the boiler. A flow diagram for a typical feed water system is shown below in Figure 11.1.

Feed water requirement is defined as follows: Boiler feed water requirement = boiler evaporation rate + boiler blowdown rate + any system steam leaks. Feed water flow = fresh makeup water rate + return condensate rate. Therefore boiler evaporation + blowdown rate + steam leaks must equal condensate return and fresh makeup rate. A water balance in the steam system would show this to be true. A steam system with "60% returns" would mean that the feed water would contain 40% makeup water and 60% condensate return water. Steam system designers should always perform a water balance early in the design process. Pumps, water treatment equipment, condensate, and feed water tank sizes can be accurately determined once the water balance is determined. More on this type of analysis in Chapter 12.

FEED WATER DEAERATION

Prior to discussing feed water system design, one should understand the principles of deaeration. Mechanical deaeration is the process of mechanically removing all dissolved gases from the feed water and is an integral part of any modern steam boiler protection program. Deaeration, coupled with other aspects of water treatment, provides the best and highest-quality feed water for boiler use. The purposes of deaeration are

Process Steam Systems: A Practical Guide for Operators, Maintainers, Designers, and Educators, Second Edition. Carey Merritt.
© 2023 John Wiley & Sons, Inc. Published 2023 by John Wiley & Sons, Inc.

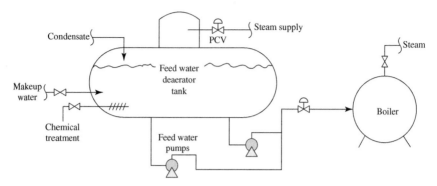

FIGURE 11.1   Feed water system concept diagram.

1. To remove oxygen, carbon dioxide, and other noncondensable gases from feed water.
2. To heat the incoming makeup water and return condensate to an optimum temperature prior to being fed back to the boiler.

The Elimination of Dissolved Gases

The most common cause of corrosion in a boiler system is from dissolved gases like oxygen, carbon dioxide, and ammonia. Of these, oxygen is the most aggressive. The importance of eliminating oxygen as a cause of pitting and iron deposition cannot be overemphasized. Even small concentrations of this gas can cause corrosion problems. The chart below in Figure 11.2 shows the solubility of oxygen versus temperature. Physical chemistry shows the solubility of dissolved gases in water is inversely proportional to the temperature [39]. Therefore, as the temperature approaches the boiling point, the dissolved oxygen concentration approaches zero.

The chart also shows that as you increase the pressure, the temperature at which full deaeration occurs is higher. This is because of the relationship between pressure and the temperature (PV/T = constant) as dictated be the ideal gas law. Good mechanical deaeration can only occur if the water temperature is kept at the boiling point that corresponds to that pressure. The temperature of the water in the storage section of a fully functioning pressurized feed water deaerator (DA) will correspond to the pressure in the vessel. Therefore, a DA kept at 5 psig steam pressure should have a water storage temperature approximately 227 F. Likewise, an open atmospheric vented feed water tank will have full deaeration when the water in that vessel reaches 212 F (at sea level).

Example 11.1

The feed water tank for a process steam system shows a feed water temperature of 150 F. What is the oxygen level of the feed water in this tank if the tank is open to the atmosphere? What would it be if the feed water were heated to 200 F?

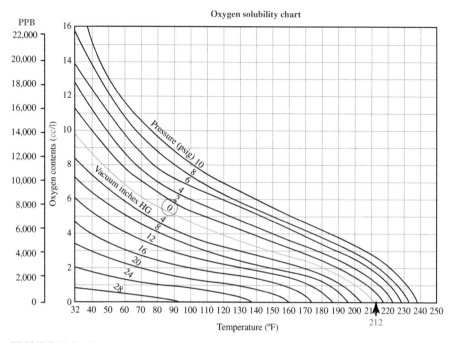

FIGURE 11.2 Oxygen solubility versus temperature. Courtesy of the Shippensburg Pump Co.

Answer: The chart in Figure 11.2 shows the oxygen level at 150 F at atmospheric pressure to be about 3 cc/lt. If the feed water temperature is heated to 200 F the oxygen levels drop to about 1 cc/lt. Heating the water an additional 50 F reduces the oxygen concentration by 66%. While 1 cc/lt is a very low amount, it is still higher than the 0.005 cc/lt recommended for complete corrosion protection.

The foregoing discussion shows the importance of proper deaeration of boiler feed water in order to prevent oxygen corrosion. Generally, complete oxygen removal cannot be attained by mechanical deaeration alone. Equipment manufacturers state that a properly operated deaerating heater can mechanically reduce the dissolved oxygen concentrations in the feed water to 0.005 cc per liter (7 ppb) and 0 free carbon dioxide. Traces of dissolved oxygen remaining in the feed water can then be chemically removed with chemical treatment, like an oxygen scavenger. Consequently, most steam system operators will add an oxygen scavenger chemical to the feed tank to remove the remaining oxygen. This will ensure complete boiler protection against oxygen corrosion. It should be noted that measuring oxygen levels below 50 ppb requires a closed flow through oxygen meter. Drawing a sample left open to atmosphere for only seconds will allow oxygen to resolubilize and give a false oxygen reading.

Makeup water introduces appreciable amounts of oxygen into the system. This is another good reason to reuse as much condensate as possible. Consequently, high makeup water steam systems will require robust deaeration. Oxygen can also enter the feed water system from the condensate return system. Possible return line sources are direct air-leakage from vacuum breakers, condensate return tanks open to atmosphere, and leakage of nondeaerated water used for condensate pump seal and/ or quench water.

FEED WATER TANKS

The feed water system's first design principle is to collect the condensate. There are two types of feed water tanks atmospheric tanks and fully pressurized feed water deaerators. In some smaller steam systems, the trapped condensate may flow directly to a feed water tank. In larger systems the feed water tank may receive water from a condensate surge tank or from several small condensate receivers.

Atmospheric tanks, similar to the one shown in Figure 11.3, are used extensively on small steam systems. The tank construction can be carbon steel or stainless steel. This type of feed water tank will look similar to a condensate collection tank, only have a larger storage volume. It is more difficult to mechanically fully deaerated feed water in an open tank. Consequently, the life cycle of these tanks will be shorter than a well operated pressurized deaerator. Because stainless steel is less susceptible to corrosion, constructing the tank with stainless steel will increase the life relative to carbon steel construction. These tanks are generally vented to atmosphere and are packaged with on/off feed water pumps. They are simple, replaceable, and much less costly than a feed water deaerator.

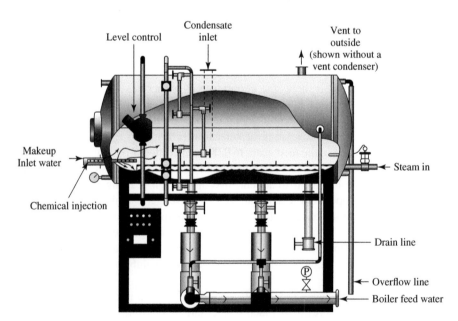

FIGURE 11.3   Typical noncode basic feed water tank.

There are a few design enhancements that can increase the life of an atmospheric tank and improve the feed water quality. Steam preheating via direct injection or the use of a steam coil will help maintain the feed water hot and somewhat deaerated. We have seen in Figure 11.2 that the hotter the water, the less oxygen it will carry. Steam preheating this way is usually performed using a tank sparge tube or injector and a direct acting temperature control valve. It should be noted that steam injection directly into the feed water will create a significant amount of noise, therefore should be taken into consideration if the feed water tank is located near a room that is occupied. The noise is a result of thousands of small steam bubbles collapsing and each creating small levels of water hammer.

When an atmospheric feed tank is used and the feed water approaches 210 F, there will be some steam vapors that will be vented off. In the case where condensate is brought back directly to the feed water tank, leaking steam traps can allow a lot of steam to be vented off from the tank. A vent condenser mounted on the tank top section can be used to condense any steam vapors and reduce the amount of freshwater makeup. Likewise a magnesium anode placed in the tank shell will provide some additional protection against tank corrosion. The tank as a minimum should be fitted with a water level sight glass, thermometer, and water treatment chemical injection ports in addition to freshwater makeup connection and a level control system.

Feed water tanks can also be pressurized ASME vessels that are very efficient at mechanically deaerating feed water. These feed water tanks are called **deaerators** and come in two types: spray type and tray type. DAs are significantly more expensive compared to an atmospheric tank but are made more rugged, last much longer, and perform much better. Feed water deaerators are used almost always on larger steam systems.

Spray-type feed water deaerator (DA), similar to the one shown in Figure 11.4, is a pressured tank capable of fully deaerating the feed water. These tanks are typically ASME section VIII code tanks with a maximum allowable working pressure of 50 psig. In this type of feed water tank, the condensate and fresh makeup water are introduced into the tank through a series of spray nozzles. In addition, steam is allowed into the tank in such a manner that will raise the temperature of the fine water droplets to the saturation temperature and fully deaerate the water. Some designs provide a recirculation loop that provides a means to continuously spray water within the tank. A steam control valve with a pressure sensor is used to keep the tank at 3–5 psig. This type of feed water tank will provide excellent feed water quality but is usually about two to three times the cost of an atmospheric tank. Life of these tanks are > 15 years if they are operated correctly. One reliability issue that can affect a spray-type DA is spray nozzles plugging. Scale or rust deposit build up in the spray nozzles can reduce the effectiveness of deaeration. Consequently, operators and maintainers should watch the DA performance and perform frequent internal inspections of the spray nozzles.

Tray-type feed water deaerator is the other type of pressurized DA that will perform to the same level as a spray-type DA. A typical Tray type DA is shown in figure 11.5. The main difference is the manner in which the feed water is deaerated. A tray-type DA uses a series of stacked vertical trays to cascade the feed water down

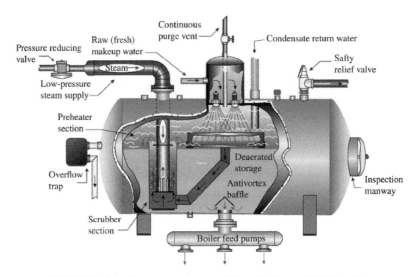

FIGURE 11.4   Spray-type deaerator.Hurst Boiler and Welding Co.

as steam is introduced horizontally or counter flow upward. The trays flatten the water flow, provide a high degree of surface area, and allow residency time in a steam pressure environment. The steam heats the water close to the steam saturation temperature, and it removes all but the very last traces of oxygen. The deaerated water then falls to the storage space below, where a steam blanket protects it from oxygen recontamination.

During normal operation, regardless of the deaerator type, the vent valve on the DA must be open to maintain a continuous plume of vented vapors and steam. If this valve is throttled closed too much, noncondensable gas will accumulate in the deaerator. This is known as air blanketing and can be remedied by increasing the vent rate. For optimum oxygen removal, the water in the storage section must be heated to within 2 degrees of the temperature of the steam at saturation conditions with the vent open enough to allow the corrosive gases to exit the tank. Measuring for dissolved oxygen in the feed water exiting the DA will likely indicate whether proper venting is occurring.

Feed Water Tank Sizing

The feed water tank must provide a means to combine the returned condensate with enough fresh makeup water to satisfy the water supply requirement of the boiler(s). Typically the storage volume of the feed water tank will be 10 min of capacity based on the maximum output mass flow of the boiler(s). One boiler horse power (bhp) will yield 34.5 lb of steam per hour under ideal conditions, which equates to 34.5 lb/h ÷ 8.34 lb/gallon = 4.14 gallon per hour of water or 0.069 gallon per minute. We can use this information to calculate how much water any size boiler will use in a minute by simply multiplying the bhp by 0.069 gpm. **A feed water tank should**

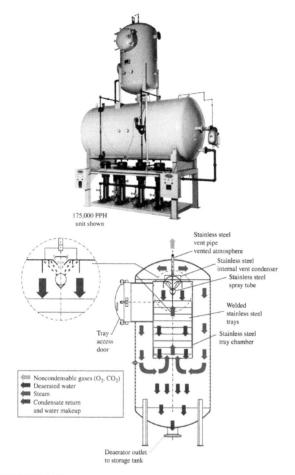

Stainless steel
vent pipe
vented atmosphere

Stainless steel
internal vent condenser

Stainless steel
spray tube

Welded
stainless steel
trays

Tray
access
door

Stainless steel
tray chamber

Noncondensable gases (O₂, CO₂)
Deaerated water
Steam
Condensate return
and water makeup

Deaerator outlet
to storage tank

FIGURE 11.5 Tray-type feed water deaerator. Bryan Boiler Co.

have a minimum working volume equal or greater than 10 minutes worth of the required boiler water volume consumed at 100% of boiler(s) output. It is highly recommended that steam systems utilizing a condensate system that does not use a surge tank, have a feed water tank sized for 15–20 min worth of boiler mass flow to accommodate surges in condensate returns.

Example 11.2
A 200 hp boiler will have a rated mass flow of 200 bhp × 34.5 lb/bhp = 6900 lb/h steam flow. Since one pound of steam equals one pound of water, then the feed water tank for this size boiler should have a capacity of at least 6900 lb/h × 1 h/60 min × 1 gal/8.34 lb = 13.8 gpm × 10 min = 138 gallon. This should be the working volume, not the full volume. Some void space is needed in the feed water tank to accommodate surges of water, especially at startup of a steam system.

When multiple boilers are being fed from a single feed water tank, the storage volume should be based on the total boiler horse power from all the serviced boilers.

FEED WATER PUMPS

The feed water system must be able to pump the hot pretreated water back to the boiler when the boiler calls for water. The feed water pump take a suction from the feed water tank and provides hot feed water directly to the boiler when the boiler's level control system senses a low operating water level. A typical feed water pumping system is shown below in Figure 11.6.

This task is not easy because the feed water in the feed water is at or very near the boiling point. The feed water pumps must have adequate capacity and **net positive suction head (NPSH)** at the feed water temperature to perform this function. NPSH is divided into both NPSHA (available) and NPSHR (required). NPSHA is the head pressure available to the pump as installed and it is calculated as such.

$$\textbf{NPSHA} = \textbf{Static head}\left(\textbf{SH}\right) + \textbf{head equivalent for water vapor}$$
$$\textbf{pressure}\left(\textbf{H}_{\textbf{VP}}\right) - \textbf{suction piping friction losses}\left(\textbf{HF}\right).$$

The static head (SH) is simply the height of water from the centerline pump suction to the top of the water line in the storage vessel. The head equivalent from vapor pressure (Hvp) is simply the difference between atmospheric pressure and the vapor pressure of the water. Think of Hvp as the resistance of water to boil expressed as an equivalent feet of head. The difference between atmospheric pressure and water vapor pressure can be converted to equivalent head for any temperature. A chart showing the relationship between Hvp and temperature for water is shown in Table 11.1.

Table 11.1 shows the hotter the water, the lower the Hvp. This makes sense as water nears the boil point, its vapor pressure increases, and its resistance to boiling

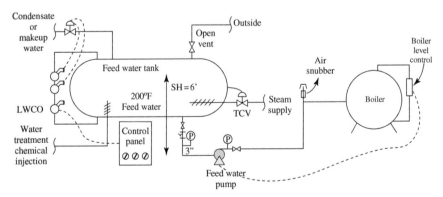

FIGURE 11.6 Typical feed water pumping system with atmospheric tank.

TABLE 11.1   **The Head Equivalent for Water Versus Temperature (Hvp)**

Temperature (°F)	Vapor Pressure (psia)	(Hvp) Equivalent ft of Water
212	14.7	0
210	14.1	1.4
208	13.7	2.6
206	13	4
204	12.5	5.1
202	12	6.3
200	11.5	7.4
190	9.3	12.5
160	4.7	23.1
120	1.7	30.1
80	0.5	32.5
40	0.1	33.1

decreases. Piping friction losses (HF) are the resistance to water flow on the suction side of the pump expressed as equivalent feet of head. If water flow to the pump is in the laminar flow range and the distance from the tank to the pump is only a few feet, then this value will be quite small and is usually ignored. If the water entering the pump is at the boiling point its vapor pressure is high, then a greater static head is required at the pump suction to ensure that the necessary NPSHA is obtained. If a pressurized feed water deaerator is functioning as designed and the pump suction piping is sized for little or no friction losses, then the NPSHA becomes essentially the static head on the pump.

The other important pressure rating for pump service is the NPSHR or net positive suction head required. NPSHR is a requirement of the pump design and can be found from the pump manufacture. **A good feed water pump selection and installation will have NPSHA at least four feet equivalent head greater than the pump NPSHR**. If the feed water is near the boiling point and not enough NPSH is available, it will begin to boil in low-pressure zone of the pump. The resultant steam is carried with the water to a high-pressure zone in the pump, where the bubbles then implode with hammerlike blows, eroding the pump and eventually destroying it. The phenomenon is called **cavitation** and is readily recognized by its typical rattlelike noise. Cavitation, which sounds like the pump is trying to pump marbles, can also occur if a pump is run at a lower discharge pressure than its optimal design. At this condition the NPSHR starts to climb steeply and it can exceed the NPSHA. Simply closing down a pump discharge valve to "fake" the pump into thinking it is pumping against a higher discharge pressure will often eliminate this cavitation. Figure 11.7 shows a typical pump curve.

ASME code for steam boilers requires the feed water equipment manufactures to use pumps that are rated for 105% the boilers safety relief valve set pressure, not necessarily the operating pressure. Consequently, a boiler with a MAWP of 150 psig and SRVs rated at 150 psig will have feed water pump rated for 157 psig discharge pressure. A pump pressure rating becomes an issue if the boiler is operated at a

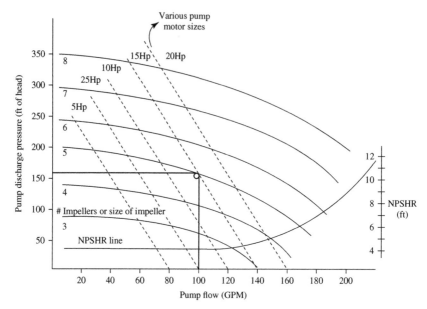

FIGURE 11.7   Typical feed water pump curve.

significantly lower pressure than the SRV set pressure. If this same 150 psig MAWP boiler is operated at a significantly lower pressure, that is, 50 psig, then the pump will operate way to the right (run out condition) on the pump curve outside of its optimal range. In cases like this, it is wise to refit the boiler with lower SRVs which would reduce the pump discharge pressure requirement and allow you to use a smaller horsepower pump. Feed water pumps are often oversized due to this discrepancy in configuration. When retrofitting is not an option, then closing down the pump discharge valve slightly to raise the pump discharge pressure will help prevent run out.

The first step in sizing the feed water pumps is to determine what type of level control system the boiler being serviced will utilize. Some of the more used configurations are shown below in Figure 11.8. The two major feed water control schemes are on/off feed water control and modulating feed water control. On/off control uses a level control system that senses a water level equal to when the boiler needs water and when the boiler is full of water. This type of level control, seen mainly with smaller boilers, can be done with an on/off feed water pump or, an on/off control valve with constant run feed water pumps. On/off feed water systems are simple and the least expensive to install; however, they inject feed water intermittently at a relatively high rate This tends to make steam production somewhat cyclic which can lead to steam pressure swings.

Modulating feed water control uses a control valve and constant run pump or a pump that has a variable frequency drive (VFD) motor. The boiler level control system converts water level to proportional electronic output signal. A control device converts this signal to a valve position or pump speed. Consequently, these type of level control systems inject water into the boiler at a rate high enough to keep the

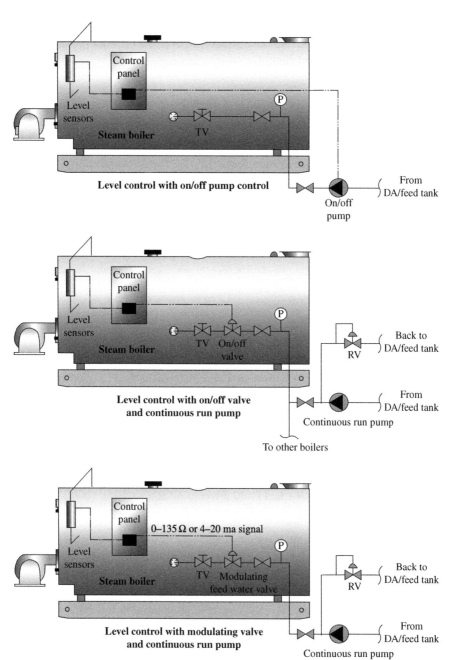

FIGURE 11.8 Feed water/level control schemes.

level within pretty tight tolerances. Generally, the larger steam systems use a modulating feed water control system. The consequence of steam pressure cycling is eliminated with this type of level control.

Feed Water Pump Sizing

It is critical to size the feed water pump correctly. Here is a good way to systematically size the pump.

    A. First, determine required pump flow rate:
- Use 0.069 gpm/bhp as water usage rate for steam boiler.
- For on/off pump use 2.5 times water usage rate.
- For modulated pumping use 1.5 times the water usage rate.

Example 11.3
*A 60 hp boiler will use 0.069 gpm/bhp × 60 bhp = 4.1 gpm. If an on/off pump application is used then you will need a pump that will pump about **10 gpm**. If the boiler has a modulating level control system, then 4.1 gpm × 1.5 = **6.2 gpm** pump capacity would be required.*

    B. Determine NPSHA:
- Need to know Total Static Head (SH)
- Need to know water temperature in the feed tank
- Need to know pump suction friction losses

Example 11.4
NPSHA calculation. From Figure 11.6 *above, we can see the static head (SH) will be 6 feet, the water temperature in this feed water tank is 200 F, which corresponds to 7.4 feet of NPSH (Hvp from* Table 11.1*) and the suction piping losses should be negligible. **Therefore the NPSHA will be 13.3 feet of head**.*

    C. Select kind of pump that best fits application
- Centrifugal pumps: high volume, high NPSHR, low discharge pressure
- Regenerative turbine: low volume, medium pressure, low NPSHR
- Multistage centrifugal: low to medium volume and pressure, very low NPSHR
- Positive displacement: low volume, high pressure, very low NPSHR

Example 11.5
*In our example, the pump selected can best be determined based on the boiler size and trim pressure. If the boiler is <100 hp and trimmed at 150 psig or lower, then a regenerative turbine type would suffice as long as the NPSHA > NPSHR + 4 feet. If the boiler is >20 hp and trimmed at <250 psig, then a **multistage centrifugal pump***

could be used, especially if we had a low NPSHA. If the pressure is above 250 psig, then a positive displacement pump might be the best choice. For larger boilers up to 150 psig, centrifugal pumps are often used.

D. Using a pump curve, select a model:
- Pump that will provide the flow at the desired pressure plus 5%
- Pump that has NPSHR > NPSHA by at least 4 feet
- Pump that has seals compatible with water temperature
- Pump that has metallurgy compatible with application
- Pump that will run at a high efficiency

Example 11.6
You have a 200 hp high pressure (150 psig MAWP) steam boiler, trimmed at 150 psig, that will use a modulating feed water level control system. A typical set up is shown below in Figure 11.9. *What is the best type and size pump to be used for this application?*

Answer: The capacity of the pump needs to be 200 bhp × 0.069 gpm/bhp × 1.5 safety factor = 20.7 gpm. Since this pump has a continuous minimum flow (4 gpm), then this flow needs to be added to pump requirements. Therefore the pump must be able to pump at least 24.7 gpm @ 150 × 105% or 158 psig at 363 feet TDH. The NPSHA is SH + Hvp – HF. From Figure 11.9 we can see SH = 10 feet, Hvp = 0, and HF is negligible, then the NPSHA is equal to the static head = 10 feet. At 158 psig discharge pressure, a multistage centrifugal pump would be a good choice. The pump curves shown in figure 11.10 below shows that pump No. 2 with seven stages would be a good selection. The pump curves also shows that the NPSHR for pump No. 1 is >8' which is too close to the 10' NPSHA you have for this application.

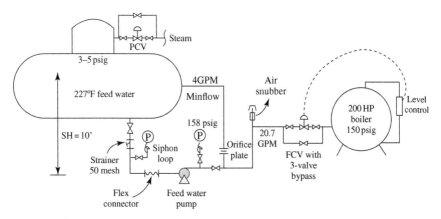

FIGURE 11.9   200 hp boiler with modulating feed water control system.

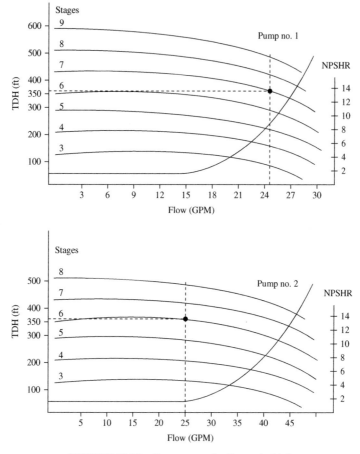

FIGURE 11.10   Pump curve for Example 11.6.

FEED WATER PIPING

The last design consideration that must be accounted for is the feed water piping. This piping like most water systems should be designed to allow for laminar flow. This is especially important for pump suction piping. Laminar flow will result from flow that has a velocity of 7 ft/s or less. Piping charts like the one found in Appendix A can be used to size pipe. A good rule of thumb is to determine discharge piping sizes based on velocities <7 ft/s and then upsize the suction piping by a ½ size or more. Table 11.2 shows some recommended flow velocities for feed water and condensate lines

For instance if the feed water pump discharge pipe size needs to be 1.5 "schedule 40 pipe, then use 2" schedule 40 pipe minimum for the pump suction pipe. Besides sizing the feed water piping correctly, equally important to include the right piping components like pressure gauges, strainers, isolation valves, snubbers, flex connectors, etc. Figure 11.9 shows a typical feed water piping scheme suitable for most feed water systems.

TABLE 11.2   Recommended Water Flow Velocities

Waterline	Velocity
Boiler feed pump suction	1.5–2.5 ft/s
Condensate pump suction	1.5–3.0 ft/s
Condensate pump discharge	3.0–7.5 ft/s
Boiler feed pump discharge	4.0–10 ft/s

Data from Watson McDaniel [18].

The frequently left out component in the feed water line is the snubber. The snubber is nothing more than a section of capped off vertical pipe that has trapped air inside. This can be a purchased piece of equipment or can be field fitted during installation. A common field installed snubber would replace a 90 degree elbow, used to change a vertical pipe run to a horizontal pipe run, with a tee fitted with a 18–24 inch section of capped off pipe. When a pump starts or a control valve slams open, water hammer can result. **A snubber acts as a shock absorber and will significantly reduce water hammer**. Any feed water piping with long piping runs should include a snubber inline between the pump and boiler. Other important feed water piping components include strainers with 50 mesh baskets in front of pumps and control valves to protect the internals.

Feed water pumps should be fitted with suction pressure and discharge gauges. The pump suction gauge should be able to display a slightly negative pressure. A good design will include a thermometer in the feed water discharge piping and pump isolation valves. For larger feed water pumps, it is recommended that a flex connector be installed on the pump suction line to eliminate pump torque stress on the feed water tank nozzle. Pressure gauges that will contact steam or feed water should be fitted with a siphon to provide an air seal to protect the gauge from pressure spikes. Reference Figure 11.11.

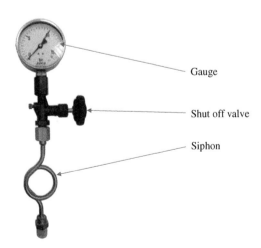

Gauge

Shut off valve

Siphon

FIGURE 11.11   Typical pressure gauge installation.

Feed Water-Surge Tank Controls

Anytime you have a boiler, feed water tank and surge tank being designed in a steam system, you must make sure the three level control system work together or you will waste a lot of water. A boiler will receive water from the feed water tank and these two systems typically work in unison without concern as described above. The issue most time is the surge tank and feed water tank level control schemes not working well together. For instance, if both the surge tank and DA have makeup water control systems, there will be a good chance the water balance will get uneven and one will eventually over flow wasting treated water.

I have found in these situations, making water up in the surge tank on low level and only allowing the feed water tank to makeup water on a low-low water level works best. This scheme will allow the surge tank to mix the cold makeup water with the hot condensate and provide a reservoir of supply water of warm water to the feed water tank. Should the surge tank not have enough water level, then the feed water tank on an off normal low-low level condition would open an emergency makeup water valve. Caution is advised when the surge tank and feed water tank/DA are from two different manufactures. Engineers need to be very specific about the desired interconnection of the level and pump controls between the three pieces of equipment.

12

STEAM SYSTEM CHEMISTRY CONTROL

Most engineers and operators understand that a good steam system monitoring and maintenance program involves some type of water treatment program. Maintaining good heat transfer efficiency and preserving the integrity of the pressure vessel boundary is the goal of a steam system water treatment program. This chapter will discuss basic water chemistry, the consequences of poor chemistry control in a steam system, and the various methods of control available today. Simply, any boiler water treatment program must control the three adverse conditions: **scaling, fouling, and corrosion**. Each is described in the following text.

BASIC WATER CHEMISTRY

Water chemistry is thought to be a mysterious complex science. The fact that we cannot see the impurities fuels this perceived mystery. If we look at each component of water, it might be easier to understand the chemistry and control methods.

Water itself needs to be defined. Water can be groundwater, surface water, city water, or reverse osmosis/demineralized (RO/DI) water. Groundwater is water taken from a well and surface water is water found in ponds, lakes, and rivers. City water is surface or groundwater that has been filtered and treated with a disinfectant like chlorine. All three of these waters contain components that can be detrimental to boiler pressure vessel life and steam quality, if not regulated within limits. RO/DI water is any water source that has had the suspended and dissolved solids removed, leaving only pure water and dissolved gases. The matrix shown in Table 12.1 shows the chemistry of the various types of water.

Process Steam Systems: A Practical Guide for Operators, Maintainers, Designers, and Educators, Second Edition. Carey Merritt.
© 2023 John Wiley & Sons, Inc. Published 2023 by John Wiley & Sons, Inc.

TABLE 12.1 Water Types and Their Chemistry [39]

Component	Groundwater	Surface Water	City Water	Reverse Osmosis/Demineralized Water	Affect on Steam System Life
Calcium (Ca)	20–50 ppm	10–25 ppm	20–50 ppm	<2 ppm	Causes scaling
Magnesium (Ma)	10–20 ppm	2–10 ppm	2–20 ppm	<2 ppm	Causes scaling
Sodium (Na)	25–125 ppm	15–80 ppm	15–125 ppm	<5 ppm	No significant affect at typical concentrations
Silica (Si)	5–10 ppm	2–10 ppm	2–10 ppm	<2 ppm	Can affect steam turbine performance
Sulfate (SO_4)	5–25 ppm	5–25 ppm	5–25 ppm	<2 ppm	Can cause foaming at high levels. Food for MIC (microbiological induced corrosion)
Dissolved oxygen (O_2)	9 ppm @ 70°F	9 ppm @ 70°F	9 ppm @ 70°F	9 ppm @ 70°F	Causes carbon steel corrosion >1 ppm
pH	6–7	7–8	7–8	5–6	pH <7 cause carbon steel corrosion; pH >12 can cause caustic embrittlement
Iron (Fe)	<2 ppm	<2 ppm	<2 ppm	<2 ppm	Negatively affect heat transfer at high concentrations
Alkalinity (HCO_3)	30–100 ppm	10–50 ppm	10–100 ppm	<5 ppm	Required buffer but will cause poor steam quality at high levels
Chloride (Ci)	30–100 ppm	10–50 ppm	10–100 ppm	<2 ppm	No significant affect on carbon steel. Causes corrosion in stainless steel
Total organic carbon (C)	1–5 ppm	5–10 ppm	1–10 ppm	<2 ppm	High levels can cause low pH and poor steam quality
Suspended solids (SS)	0–10 ppm	10–50 ppm	0–5 ppm	<2 ppm	Can cause fouling at high levels
Total dissolved solids (TDS)	100–200 ppm	50–100 ppm	50–200 ppm	<10 ppm	High levels cause carbon steel corrosion and poor steam quality
Dissolved carbon dioxide (CO_2)	20–40 ppm	2–10 ppm	2–10 ppm	20–70 ppm	Lowers water pH
Chlorine	ND	ND	0.5–2 ppm	ND	Corrosive to CS and water softener resin

Water is comprised of five distinct components.

1. Pure water (H_2O): It has a pH of 5–6 and if untreated is very corrosive to carbon steel.
2. Ionic components (Ca, Na, Mg, SO_4, HCO_3...etc.) are dissolved ions in the water and increase the water's ability to conduct electricity. Higher ionic concentrations generally mean high general corrosion potential. However, some of these ions are good and essential for pressure vessel life.
3. Dissolved gases (CO_2, O_2, and N_2): Some of the gases like oxygen and carbon dioxide are corrosive and become less soluble in water with increasing temperature.
4. Organics (soaps, solvents, vegetation...etc.) can be soluble or insoluble. They can break down and affect water pH and surface tension. In high concentrations, increased corrosiveness and poor steam quality will occur in an operating boiler.
5. Inert matter (dirt, silica, corrosion products...etc.) can plate out on the pressure vessel causing fouling at high concentrations and retard heat transfer.

All of these water components can be detrimental to boiler pressure vessel life and steam quality if not maintained at threshold levels or rendered harmless via proper treatment. A good water treatment program will use both chemical treatments (internal treatment) and nonchemical methods (external treatment) to prevent adverse conditions.

Generally, one can consider what is found in water to be good for animal life is not good for a boiler system. For instance, calcium and magnesium, two elements that health professionals say we need for strong bones, are the leading contributor to boiler **scale** formation. Similar concepts can be said for oxygen and iron. We need them, but the steam system does not. There are three distinct conditions, scaling, fouling and corrosion, that must be controlled to ensure long life, reliable, and efficient operation of any steam system.

SCALE CONTROL

Scale is that light-colored layer that forms on the waterside of a heated metal surface. This layer can be hard as dried paint or soft as jelly. Either condition tends to significantly reduce the heat transfer (efficiency) of heat through the metal pressure boundary. Calcium and magnesium commonly referred to as hardness salts, are found in almost all natural water sources, including city water. **Hardness salts become less soluble as the temperature increases**. Therefore, in a boiler on the waterside of the metal surfaces where the hot burner gases transfer energy, these hardness salts tend to precipitate out. The precipitation forms a layer which acts as an insulator on the metal. Boilers designed with high heat flux furnaces are very susceptible to scaling if the water chemistry is not controlled.

Control of scale formation can be accomplished by either removal of the calcium and magnesium from the supply water or by adding a chemical to "tie up" the hardness salts and prevent precipitation. A newer technology based on using a magnetic field to change the polarity of the hardness ions, thus preventing scale formation, has been introduced to the market in recent years. On many boiler systems, especially larger ones, a combination of control methods is used. Removal of hardness from water can be accomplished by several **external control** technologies; however, the most popular are softening, deionization, and reverse osmosis. Softening and deionization are methods that use the principles of ion exchange. Consequently, these technologies require regeneration of the media once the exchange capacity is "spent."

Softening is the process of replacing hardness ions with sodium or hydrogen ions. An ion exchange media is used to selectively perform this function. The ion exchange media or resin has millions of small pores that are used to bind the hardness ions. It is important to realize that each resin bead has a finite amount of hardness binding capacity and will therefore need to be regenerated periodically. Regeneration simply floods the resin with sodium ions (salt) or acid to reactivate the pores. Hardness is removed by passing makeup water through the softener until the resin capacity is diminished to a level that requires regeneration. Softeners can regenerate based on time of day, flow of gallons throughput, or manually activated. A typical water soft-ener is shown in figure 12.1. Generally, the water treatment folks will sample makeup water, determine hardness level, and set up the regeneration cycle based on makeup water usage rate. Water softeners are sized based on the hardness removal capacity. The formula below can be used to size the softener. Boiler size (BHP) × 0.069 gpm/BHP × 60 min/h × 24 h/day × % makeup × average load factor × makeup water hardness (grains/gallon or gpg) = daily softener capacity.

Example 12.1
A 200 hp boiler with 50% makeup water with a water quality of 14 grains of hardness/gallon and a load factor of 75% will require a softener sized for 200 bhp × 0.069 gpm/bhp × 60 min/h × 24 h/day × 50% × 75% × 17 gpg = 104,328 grains/day capacity.

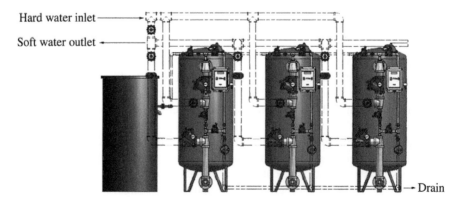

Hard water inlet

Soft water outlet

Drain

FIGURE 12.1   Water softener systems. Marlo Inc.

Deionization (DI) uses demineralizing resin to remove all the dissolved solids from the water. Unlike softening, which selectively removes calcium and magnesium, DI systems remove all components. The DI resin which is stored in vessels removes the water impurities and replaces them with H^+ and OH^- ions which reform to make more H_2O or water. DI resin like softening resin requires periodic regeneration. DI systems are used extensively in purifying makeup water and for condensate polishing in larger power generating steam systems. One issue with DI systems is they remove the good components of water along with the bad components. Some downstream water treatment is required to adjust the system pH.

Reverse osmosis uses the principles of membrane filtration and pressure to separate the pure water from its impurities. A typical RO system is shown in Figure 12.2. Both RO and DI water treatment methods will produce nearly pure water. The membranes used in RO systems have pores sizes to allow only the water molecules to pass through them. The dissolved solids and some of the supply water remain. The pure water is called permeate, and the concentrated brine waste water is called centrate. A good operating RO system will recover at least 75% of the supply water as permeate. RO systems are widely used in small and medium size steam systems. It should be noted that there are two consequences that must be addressed when using an RO system: supply water pretreatment and postwater treatment. Supply water to a RO should be prefiltered to remove gross suspended solids. If city water is the supply water source, then the free chlorine should be removed or rendered harmless via carbon filtration or sodium sulfite addition. Chlorine will ruin RO membranes even at very low concentrations over time. Likewise, the post RO permeate (and post DI) may have a pH <6 is corrosive to carbon steel and should be neutralized to 7.5 or 8 prior to allowance into the boiler feed water system.

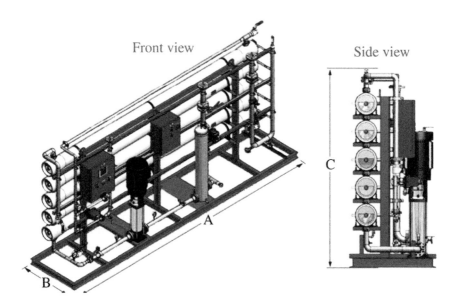

FIGURE 12.2 Reverse osmosis system. Marlo Inc.

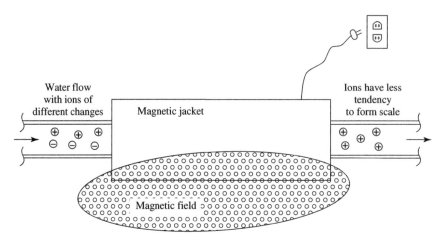

FIGURE 12.3   Magnetic field-scale prevention concept.

The third technology used on boiler systems uses a **magnetic field** to change the polarity of the hardness ions. This method of control relies on performance of a magnetic field and the theory that when you change the polarity of the hardness ions scale precipitation will not occur. This technology, which is relatively new, uses a magnet strapped to the feed water piping and an electrical current is supplied to the magnet. The device creates a magnetic field around the pipe which affects the ions in the feed water as it passes through that section of pipe. Although this technology has been around for years, I could not find a suitable commercial example of its performance as measured over time. The concept is shown in Figure 12.3.

FOULING CONTROL

The second detrimental condition that will lead to steam system failure is **fouling**. Fouling is the precipitation of corrosion products (rust), scale, and dirt in low-flow areas. The most prevalent place fouling occurs is on the bottom head or mud drum of the boiler. Here the buildup of solids can inhibit the heat transfer so much that severe stress and cracking can occur. Fouling is a consequence of maintaining high suspended solids in the boiler water. High suspended solids are caused by the precipitation of dissolved solids or carryover of corrosion products from the feed water system. In either case, the solids must be purged frequently from the boiler to prevent fouling. Iron is carried into the boiler in various forms of chemical composition and physical state. Most of the iron found in the boiler enters as iron oxide or hydroxide. Any soluble iron in the feed water is converted to the insoluble hydroxide when exposed to the high alkalinity and temperature in the boiler.

Corrosion products in the feed water can cycle up and concentrate in the boiler. As a result, deposition takes place on internal surfaces, particularly in high heat transfer areas, where it can be least tolerated. Metallic deposits act as insulators, which can cause local overheating and failure. Deposits can also restrict boiler water circulation. Reduced circulation can contribute to overheating, film boiling, and accelerated deposition. Migratory iron, migratory copper, and other contaminants such as calcium, magnesium, and silica must be conditioned within the boiler itself. Excursions of calcium, magnesium, and silica can create deposition problems within the feed water train. Minimization of these contaminants prior to the feed water train is the most successful way of dealing with this problem. We have boiler water total dissolved solid (TDS) levels and makeup water total suspended solid (TSS) levels to help mitigate fouling potential. Boiler blowdown will keep the TDS level in check. Makeup water mechanical filtration or RO/DI is used to eliminate TSS entering the steam cycle. A section later in this chapter is dedicated to boiler blowdown.

CORROSION CONTROL

The last detrimental condition is **corrosion**. Localized attack on metal can result in forced shutdown and equipment failure. Because boiler systems are constructed primarily of carbon steel and the heat transfer medium is water, the potential for corrosion is high. Boiler water pH and ionic strength influence corrosion rates. An exapmple of iron corrosion is shown in Figure 12.4. To maintain pressure vessel integrity over several years, corrosion must be kept near 5MPY (0.005 inches/year). Iron corrosion compounds are divided roughly into two types: red iron oxide (Fe_2O_3) and black magnetite oxide (Fe_3O_4). The red oxide (hematite) is formed under oxidizing conditions that exist, for example, in the low pH condensate system or in a boiler that is out of service and subjected to water with high amounts of dissolved oxygen. The black oxides (magnetite) are formed under reducing conditions that typically exist in a properly operated boiler. Magnetite is the desired type of iron formed because it will deter further iron corrosion. A magnetite layer on a boiler pressure vessel is shown in Figure 12.5. If the system internals show a red to rust color then you have hematite being formed and the metal is not protected.

Corrosion can be general corrosion or localized corrosion. General corrosion also referred to as rusting, occurs over an area with no discernible discrete pattern. General corrosion occurs any time water is in contact with metal in the presence of dissolved oxygen or low pH water. Therefore it is imperative to maintain the boiler water dissolved oxygen at very low levels and keep the pH level within specific levels. Dissolved oxygen control is accomplished by good feed water deaeration and by the use of oxygen scavenging water treatment chemicals, like catalyzed sulfites. Since oxygen solubility in water goes down as you raise the water temperature, feed water preheating is a relatively inexpensive way to reduce feed water dissolved oxygen. This oxygen temperature relationship is shown in Figure 12.6. PH control relative to general corrosion is very important when using RO/DI water in a carbon steel boiler. RO/DI feed water must be neutralized with bicarbonate (alkalinity) prior to injection in the boiler of the boiler pressure vessel will corrode continuously. Likewise, condensate can have a low pH which can cause general corrosion in the condensate and feed water systems.

FIGURE 12.4   Typical iron corrosion of fan impellor in contact with water vapors high in oxygen content.

Other types of corrosion include localized corrosion and stress corrosion. Localized corrosion is most likely seen as oxygen pitting and will occur at water lines in the boiler and DA, or under sediment and show as pits in the metal. Stress corrosion can occur at any stress point like a weld or pipe bend. Here the stress point will corrode quickly if a corrosive environment is also found. Environmental conditions that can cause this type of stress corrosion include thermal shock, high pH/chemical levels, or high chloride concentrations. Frequently, stress corrosion associated with thermal

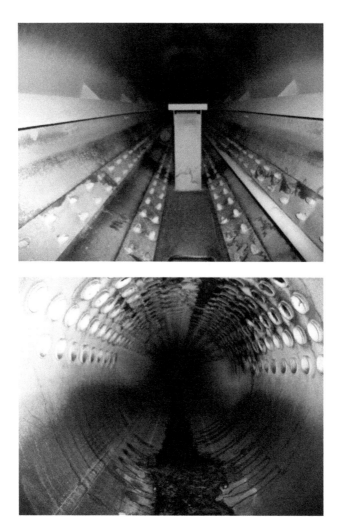

FIGURE 12.5 Water tube boiler steam and mud drum showing protective magnetite layer. Courtesy of Sunmarks, LLC.

shock is called fatigue corrosion, and the stress corrosion associated with high pH is called caustic embrittlement. Likewise, the corrosion associated with high chloride concentrations is called intergranular stress corrosion cracking (IGSCC). This type of corrosion is associated with stainless steel metal only. This is the reason we have feed water oxygen, pH, temperature, and chloride limits to help mitigate this corrosion potential.

Some other water chemistry parameters must be controlled to ensure good steam quality is maintained from the boiler at all times. High alkalinity, organics like oil

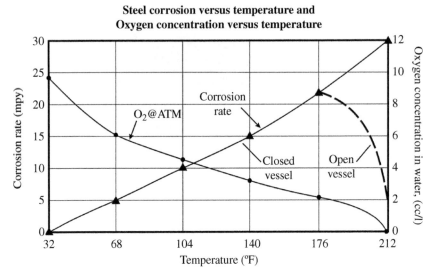

FIGURE 12.6 Corrosion/temperature and oxygen relationship. Data from MTI Pub. No. 45 [40].

and soaps, and total dissolved solids will increase the boiler water surface tension and cause foaming in the boiler. Excessive foaming will create carryover and poor steam quality. We therefore have additional limits on these parameters. The condensate from steam will contain dissolved carbon dioxide called carbonic acid, with a low pH and will be corrosive to carbon steel. Neutralizing amines are often added to the boiler or to the steam lines to effectively neutralize the carbonic acid that forms when steam condenses.

The major gaseous corrosive contaminants are removed through proper chemical and mechanical deaeration. Catalyzed sodium sulfite is a popular water treatment chemical used to remove trace amounts of dissolved oxygen and prevent further corrosion. Likewise alkalinity builder-type chemicals are used to maintain proper boiler water pH. More on chemical treatment is explained later in this chapter.

BOILER BLOWDOWN

Even with the best pretreatment programs, boiler feed water often contains some degree of impurities, such as suspended and dissolved solids. The impurities can remain and accumulate inside the boiler as the boiler operation continues. The increasing concentration of dissolved solids may lead to carryover of boiler water into the steam, causing damage to piping, steam traps, and even process equipment. The increasing concentration of suspended solids can form sludge, which impairs boiler efficiency and heat transfer capability.

To avoid boiler problems, water must be periodically discharged or "blown down" from the boiler to control the concentrations of suspended and total dissolved solids in the boiler. Surface water blowdown is often done continuously to reduce the level of dissolved solids, and bottom blowdown is performed periodically to remove sludge from the bottom of the boiler. When the steam bubble is released to the boiler steam space, the bubble will leave a slightly higher concentration of dissolved solids behind. Therefore, at the water surface, there will always be a slightly higher concentration of dissolved solids, making surface blowdown a very efficient method of purging the boiler. The importance of routine boiler blowdown is often overlooked. Once per day bottom blowdown schedule should be used. Remember, the blowdown water has the same temperature and pressure as the boiler water; therefore, the blowdown water heat can be recovered and used to heat makeup water.

Best Operating Practices for Boiler Blowdown

As mentioned earlier, insufficient blowdown may cause carryover of boiler water into the steam or the formation of deposits on boiler tubes. Excessive blowdown wastes energy, water, and treatment chemicals. The blowdown amount required is a function of boiler type, steam pressure, chemical treatment program, and feed water quality. The optimum blowdown amount is typically calculated and controlled by measuring the conductivity of the boiler feed water. Conductivity is a viable indicator of the overall total dissolved solid concentrations. The proper blowdown rate can be calculated as a function of the boilers evaporation rate. The American Society of Mechanical Engineers (ASME) has developed a best operating practices manual for boiler blowdown. The recommended practices are described in Section VI and VII of the ASME *Boiler and Pressure Vessel Code*. You can identify energy-saving opportunities by comparing your blowdown and makeup water treatment practices with the ASME practices. The ASME: *Boiler and Pressure Vessel Code* can be ordered through the ASME website.

Automatic Versus Manual Blowdown Controls

There are two types of boiler blowdowns: manual and automatic. Boiler plants using manual blowdown must check boiler water samples frequently and adjust blowdown accordingly. With manual boiler blowdown control, operators are delayed in knowing when to conduct blowdown or for how long. They cannot immediately respond to the changes in feed water conditions or variations in steam demand. Manual bottom blowdown is generally an acceptable practice for process boilers that operate at a relatively constant rate and use a continuous surface blowdown scheme.

A conductivity-based automatic blowdown control constantly monitors boiler water conductivity and adjusts the blowdown rate accordingly to maintain the desired water chemistry. A probe measures the conductivity and provides feedback to the controller driving a modulating blowdown valve. An automatic blowdown control can keep the blowdown rate uniformly close to the maximum allowable dissolved solids level, while minimizing blowdown and reducing energy losses. Automatic blowdown control is generally a good practice for larger process boilers employing surface blowdown control.

A compromise between an automatic conductivity-based blowdown system and a pure manual system is a timed-based automatic blowdown system. In this system, a programmed timer opens a blowdown valve for a specified period. By regulating flow via a throttle valve, frequency of the blowdown and the duration of the blowdown, an operator can maintain the solids level in the boiler. This type of system is ideal for a small system with a relatively constant steam load. The three types of surface blowdown systems are shown in Figure 8.6.

Determining Blowdown Rate

Boiler blowdown (BD) rate can be estimated if you know the effective evaporation rate (EER) and using the following formula. BD = EER/1 − cycles of concentration. Cycles of concentration (COC) are the ratio of dissolved solids in the boiler versus what is in the makeup water supply. A boiler operated with 2000 ppm dissolved solids that has a makeup water supply containing 200 ppm dissolved solids is said to have water with 2000/200 or 10 cycles of concentration. Likewise, the effective evaporation rate is related to the percent return as follows. EER = abhp × 0.069 gpm/bhp × 1- the fraction of condensate return, where abhp is the actual boiler output for a given period. Therefore a 100 hp boiler operating 50% of the time at full load, or 100% of the time at 50% load will have an actual boiler output, abhp = 50.

Example 12.2

A 200 hp boiler operating at 10 cycles of concentration will require what blowdown rate under these conditions:

1. *At rated output and 25% condensate return*
2. *At rated output and 75% condensate returns*
3. *At 50% output and 50% condensate returns*

Answer:

1. *At rated output and 25% returns, the boiler will have an effective evaporation rate = 200 abhp × 0.069 gpm/bhp × (1–0.25) or 10.35 gpm. At 10 cycles of concentration the blowdown rate can be calculated as such. BD = EER/1 − cycles of concentration = 10.35/1 – 10 = **1.15 gallons per minute**.*
2. *If this same boiler were to return 75% of the condensate, then the EER becomes 200 abhp × 0.069 gpm/bhp × (1–0.75) = 3.45 gpm. At 10 cycles, the blowdown rate is 3.45/9 = **0.38 gallons per minute**. Obviously, the condensate return percent has a tremendous effect on the blowdown rate.*
3. *At 50% output and 80% return rate, the EER becomes (200 × 0.5) abhp × 0.69 gpm/bhp × (1 – 0.8) = 1.38 gpm. The blowdown rate then is 1.38/9 = **0.15 gpm**.*

Increasing the blowdown rate beyond a certain level will eventually gain little benefit. Each boiler owner or water treatment specialist should prepare a graph like the one below in Figure 12.7 to show where the point of diminishing return is for cycles of concentration. The graph below shows the blowdown rate versus cycles of

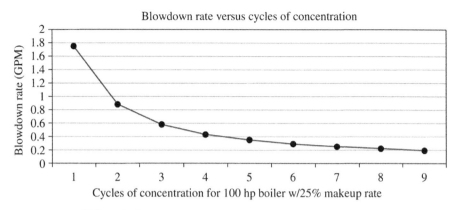

FIGURE 12.7   Cycles of concentration versus blowdown rate.

concentration for a 100 hp boiler operating with 75% returns. Figure 12.7 shows about six cycles is the point it does not make any more sense to cycle up the boiler. This curve will look similar for each steam system; however, it will shift right or left depending on the return rate.

There are some rules of thumb that can be used to estimate the blowdown rate. The rates below will yield about six cycles of concentration in the boiler. They are

1. **At 100% condensate returns, use 18 gal/h per 100 abhp.**
2. **At 50% condensate returns, use 45 gal/h per 100 abhp.**
3. **At 0% condensate returns, use 72 gal/h per 100 abhp.**

It is pretty easy to see that with a low condensate return rate, you would need to blowdown a significant amount of boiler water to stay at six cycles of concentration.

CHEMICAL FEED SYSTEMS

Most steam system applications cannot control scaling, fouling, and corrosion potential by external methods alone. Consequently, the use of internal methods, that is, water treatment chemicals are widely used. Introduction of water treatment chemicals to the steam system is a very important aspect of design. A typical chemical feed system, as shown in Figure 12.8, will consist of a tank, pump, feed line, and chemical quill. In addition, secondary containment for the tank is a very good practice and should be employed to meet 110% of the tank volume whenever possible. The design criteria that is unique to each application is how this chemical feed system is controlled.

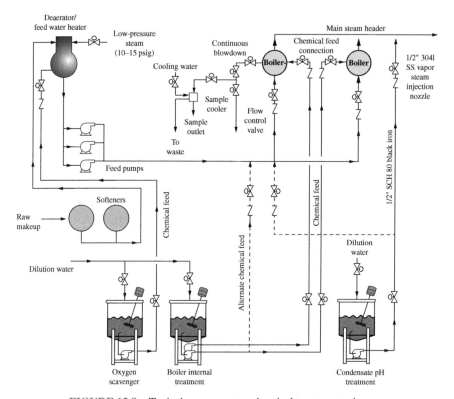

FIGURE 12.8   Typical steam system chemical treatment scheme.

There are many different water treatment schemes used today. However, almost all of them involve oxygen scavenging, pH control, fouling, and scale control chemicals.

Oxygen scavenger chemical feed systems usually add the chemical directly to the feed water tank. The chemical feed pump is usually, logic tied to a makeup water meter, feed water pump operation, feed water control valve, or be manually operated based on the boiler chemical level. The manual method should only be used for steam systems that operate under constant load conditions. Grab sampling of the boiler water for dissolved oxygen or sulfite residual is the preferred method for determining treatment effectiveness. A popular oxygen scavenging chemical is catalyzed sodium sulfite.

Boiler pH/alkalinity control chemical feed systems will almost always add the chemicals (either dilute caustic or bicarbonate) directly to the feed water line between the feed water tank and the boiler. The one exception is when the makeup water is RO/DI water, then the pH adjustment chemicals are added after the RO/DI upstream of the feed water tank. Quality control of the effectiveness of this treatment system is done by manually sampling the boiler water pH or continuously monitoring via a boiler waterside stream pH probe. Various hydroxides or carbonate salts are often used for pH and alkalinity control.

Condensate pH control is accomplished by adding a carbon dioxide scavenger directly to the boiler or steam line. These water treatment chemicals are usually amine type chemicals that will neutralize the carbonic acid formed when steam condenses in the condensate system. A steam flow meter or a feed water flow meter can provide the signal to regulate flow of this chemical into the steam system. Measuring condensate pH is the best quality control for the effectiveness of this treatment scheme.

Boiler fouling and scaling control chemicals are used to tie up the hardness ions or used to keep the suspended solids from settling out in the boiler. Polymers and or chelating agents are frequently used for this purpose. Injection of the chemical can be in the feed water tank, in the feed water line prior to the boiler, or directly into the boiler. Measuring the active chemical in the boiler either by manual sampling or by continuous monitoring is an acceptable way to monitor the chemical feed. A side stream scale/fouling monitor is the best way to evaluate the long-term effectiveness of this program.

Steam system operators should consult the boiler manufactures and water treatment professionals to develop a comprehensive chemical treatment scheme.

CHEMISTRY LIMITS

The American boilers Manufactures Association (ABMA) has set some general boiler water chemistry limits and they are shown in Table 12.2. These limits are guidelines only and in many cases do not reflect recommended chemistry limits required by some boiler manufactures. The boiler's operations and maintenance manual should provide chemistry limits that must be maintained for warrantee purposes. Consequently, each owner and operator should consult one of the several available water treatment company professionals to set up a program that ensures the three detrimental conditions are controlled.

STEAM SYSTEM METALLURGY

Carbon steel or stainless steel. The ASME regulations provide some guidance. Stainless steel cannot be used in any fuel-fired steam boiler pressure vessel construction. SS pressure vessel are limited to electric boilers and unfired steam generators. Standard steam pressure vessel are constructed from various grade of carbon steel. Additional steam system construction recommendations are shown in Table 12.3.

TABLE 12.2 Boiler Water Chemistry Limits

Boiler Feed Water Drum Pressure (psig)	Iron (ppm Fe)	Copper (ppm Fe)	Boiler Water Total Hardness (ppm CaCO₃)	Silica (ppm SiCO₂)	Total Alkalinity** (ppm CaCO₃)	Specific Conductance (micromhos/cm)
0–300	0.100	0.050	0.300	150	700*	7000
301–450	0.050	0.025	0.300	90	600*	6000
451–600	0.030	0.020	0.200	40	500*	5000
601–750	0.025	0.020	0.200	30	400*	4000
751–900	0.020	0.015	0.100	20	300*	3000
901–1000	0.020	0.015	0.050	8	200*	2000
1001–1500	0.010	0.010	0.000	2	0***	150
1501–2000	0.010	0.010	0.000	–1	0***	100

Data from ABMA (American Boiler Manufactures Association).

Source: ASME Research Committee on Water in Thermal Power Systems.

*Alkalinity not to exceed 10% of specific conductance.

**Minimum level of OH alkalinity in boilers below 1000 psi must be individually specified with regard to silica solubility and other components of internal treatment.

***Zero in these cases refers to free sodium or potassium hydroxide alkalinity. Some small variable amount of total alkalinity will be present and measureable with the assumed congruent control or volatile treatment employed at these high-pressure ranges.

TABLE 12.3 Steam System Recommended Metallurgy

	Carbon Steel	Stainless Steel	Black Iron Pipe	Polypropylene Tubing	Copper	PVC	AL29-4C
Boiler PV, fuel fired	x						
Boiler PV, electric std steam	x						
Boiler PV, electric clean steam		304/316L					
Unfired steam generator, std steam	x						
Unfired steam generator, clean steam		304/316L					
Steam header and branch mains	Sched. 40						
Condensate return*	Sched. 80	304					
Feed water piping	Sched. 40						
Blowdown piping	Sched. 40						
Raw makeup water piping	Sched. 40						
Softened water piping		304					
Chemical feed lines		304		x			
Natural gas line			x				
#2 oil lines			x				
DA. Blowdown/flash tank vents	Sched. 40						
Boiler venting					Type L	Sched. 40	x

13

MECHANICAL ROOM CONSIDERATIONS

The boiler room is the place where steam is generated and sent out to the rest of the plant. An effective boiler room design uses good steam generator selection, adequate combustion air supply, proper combustion flue gas venting, provisions to remove excessive room heat, and provide the required utilities. This chapter will look at each of these parameters and provide guidance to help ensure the boiler room is safe, comfortable, and set up to perform its function.

Four criteria should be considered when designing a boiler installation to meet the application needs. The criteria are

1. Codes and standards requirements
2. Steam load and demand profile
3. Steam system performance considerations
4. Environmental considerations

CODES AND STANDARDS

There are a number of codes and standards, laws, and regulations covering boilers and related equipment that should be considered when designing a system. Regulatory requirements are dictated by a variety of sources and are all focused primarily on safety. Here are some key rules to consider:

- The boiler industry is tightly regulated by the American Society of Mechanical Engineers (ASME) and the ASME Codes, which govern boiler pressure vessel

Process Steam Systems: A Practical Guide for Operators, Maintainers, Designers, and Educators, Second Edition. Carey Merritt.
© 2023 John Wiley & Sons, Inc. Published 2023 by John Wiley & Sons, Inc.

- design, inspection, and quality assurance. The boiler's pressure vessel must have an ASME stamp. Deaerators, economizers, and other pressure vessels must also be ASME stamped. The stamp indicates that the vessel is designed to ASME standards by an ASME qualified engineer. Some of the ASME stamp designations are shown in Appendix A. Figure 13.1 shows a sample ASME product certification nameplate.

- ASME also provides piping requirements. ASME B31.1 covers power piping, and ASME B31.3 covers process piping. Guidance in both of these codes are frequently used in designing process steam systems. Lastly, ASME specifies boiler external piping or BEP fabrication and inspection criteria. (56) BEP boundary is considered all nozzles flanged or threaded from the pressure vessel to the first or second isolation valve. See Figure 13.2 for BEP boundary definition.

- The insurance company insuring the facility or boiler may dictate additional requirements. Boiler manufacturers provide special boiler trim according to the requirements of the major insurance companies. Special boiler trim items usually pertain to added safety controls. Factory Mutual (FM) is a well-known insurance company standard.

- A UL, CUL listing, or CSA, CGA, or CSD-1 certification requirements may be required. In addition, state, local, or provincial authorities may require specific boiler controls or boiler registration and identification placards.

- National Fire Protection Association (NFPA), specifically NFPA 85, has been the standard for providing guidance in the prevention of boiler explosions. The regulation specifies burner management system logic and prepurge time requirements [42]. In addition, NFPA 54 and 31 set guidelines for venting gas- and oil-fired equipment.

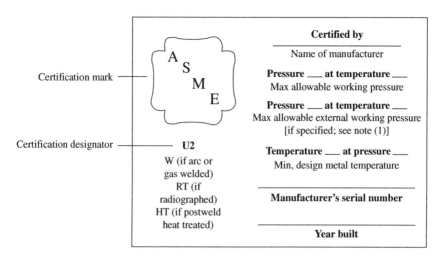

FIGURE 13.1 Sample ASME product certification nameplate. Reproduced from ASME 2013 by permission of the American Society of Mechanical Engineers [41].

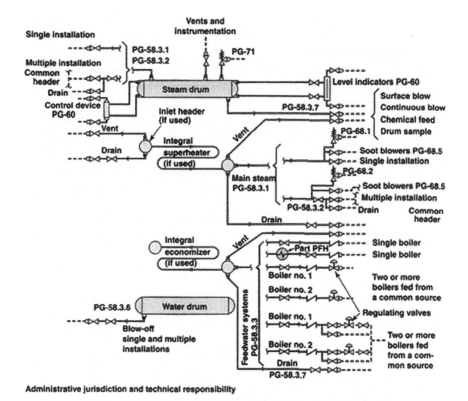

FIGURE 13.2   ASME Section I jurisdictional limits and clarifications for jurisdiction over boiler external piping (BEP) and nonboiler external piping (NBEP). (Figure PG-58.3.1, ASME Section I.) [56].

- American Society of Heating, Refrigerating, and Air Conditioning Engineers (ASHRAE) have developed standards for designing steam and hydronic heating systems. Many of the guidelines in these standards can be applied to process steam systems (i.e., boiler combustion air requirements).

- Some industries, such as food processing, brewing, or pharmaceuticals, may also have additional regulations on steam quality that will have an impact on the boiler system. A common food grade steam quality regulation is called 3A food grade quality. Health care facilities generally require an N+1 operational requirement. This require redundancy for the major steam system equipment.

- Most areas have established a maximum temperature at which water can be discharged to the sewer. In this case, a blowdown separator after cooler is required. Sites with pollution discharge permits (i.e., NPDES or SPDES) need to take into consideration the potential waste water contaminates if the water is to be discharged to the environment.

• Most state, local, or provincial authorities require a permit to install and/or operate a boiler. Additional restrictions may apply in nonattainment areas where air quality does not meet the national ambient air quality standards and emission regulations are more stringent.

• A full-time boiler operator or attendant may be required. This requirement depends on local code specifying the requirement based on the boiler's size, pressure, heating surface, or volume of water. Boilers can be selected which minimize or eliminate the attendant requirement, either by installing multiple small boilers or installing special monitoring equipment.

• Most states or provinces require minimally an annual internal boiler inspection. There may be other requirements on piping as well. The National Board of Boiler and Pressure Vessel Inspectors website www.nationalboard.com is a good place to find inspection information.

In all cases the design authority for any new or modified steam system should do a complete code review to ensure the boiler room will meet code after it has been built or retrofitted.

STEAM LOAD PROFILE

System load profile is a measure of steam demand (at a specific pressure and temperature) at any given time period when steam is required. It would be nearly impossible to size and select the right type boiler(s) without knowing the system load profile. Essentially the load profile can be defined as the system startup steam demand, the maximum and minimum steady-state steam demand, and any level of steam demand in between these conditions. Steam demand can be compared to driving your car on a trip. When you drive your car, (i.e., in city traffic), your car engine has various loads placed on it and has to adjust its output. A steam system is similar, as loads come online and go offline the demand for steam changes. The proper boiler and boiler room ancillaries need to be selected to meet the various steam load demand. The system load profile provides the following information:

• The maximum steam flow periods, including cold startup and 100% load requirement
• The minimum steam flow periods
• The rate and frequency of steam flow changes
• The idle periods
• The pressure/temperature tolerances that are required to support the process

A graphical display of some typical steam load demand profiles are shown below in Figure 13.3.

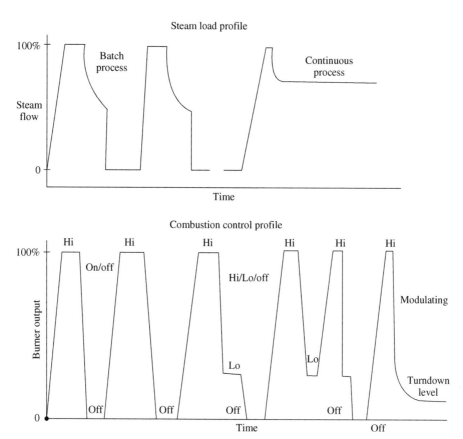

FIGURE 13.3   Steam load and combustion control profiles.

A process load is usually a high-pressure steam load pertaining to a manufacturing operation, where heat from steam is used in the process. A process load can be continuous or batch. In a continuous load, the demand is fairly constant. In a batch load steam is used in short-term demands. Continuous steam demand can be found in a refinery reboiler where a constant supply is required to continuously distill the product. A batch-type steam load profile can be found in a process tank heating application where the tank is filled, heated, and then emptied as part of a batch process.

A batch-type process load can have a very large instantaneous demand that can be several times larger than the rating of the boiler. For example, an autoclave can consume a large amount of steam simply to fill and pressurize the vessel. When designing a boiler room for a process load with high instantaneous demand, the right-type boiler needs to be selected and extra room for steam accumulators maybe required to help respond to these high steam demand periods. **High instantaneous steam loads can better be met with boilers that contain a high amount of water**. When the steam demand exceeds the output of the boiler, the boiler will depressurize and cause flash steam production. The higher the amount of boiler water will create higher amounts of flash steam. The use of steam accumulators to help mitigate the consequences of high steam demand is described in detail in Chapter 9. In addition to high load demands, a steam boiler must be capable of handling the minimum load. Boiler turndown, or the minimum steam production output, is a very important design criteria but many times overlooked. **If the boiler turndown is not adequate, then excessive boiler cycling will occur at low demand periods**. When several smaller boilers (versus one large boiler) are used to maintain the steam load, the overall turndown of the system can be greatly enhanced. Installing a 400 hp boiler with a true 8:1 turndown, will allow the boiler to meet load demands as low as 50 hp. If the 400 hp was replaced by two 200 hp boilers with the same turndown, then these boilers can satisfy a load demand down to 25 hp. This is a major consideration for replacing large boilers with smaller units. Furthermore, with today's PLC-based controls, a Building Management System (BMS) or Digital Process Computer (DCS) can provide very tight control over the boiler room output and match the system load by operating one or more boilers in an extremely efficient manner.

STEAM SYSTEM PERFORMANCE CONSIDERATIONS

Generally, the overall steam system capacity is the main performance consideration in a steam system. While capacity is very important, other performance considerations include startup to steam time, emissions control, use of economizers, local or remote control, redundancy, and steam quality. All of these considerations should be accounted for during the design phase of the boiler room.

Steam system capacity. One would think if you add up all the steam load requirements, this would equal the total amount of steam needed. While this is true, care must be taken extrapolating that data back to sizing the boiler. Boiler manufactures rate their boilers output based on 212 F feed water and 0 psig. These conditions are ideal and never seen in the field. In reality, energy that otherwise would be used for steam generation, is needed to bring feed water temperature up to the boiling point in a pressure vessel before any steam is generated. Furthermore, there will be system losses in the piping, boiler blowdown water discharges, and entrained water in the steam. These combined losses are sometimes referred to as "pick up and piping losses" and cannot be ignored. **A good rule of thumb for engineers is to multiply the total steam load by 133% to get the system gross load** [43].

Example 13.1
A steam system for a new process is being designed to provide steam to a heat exchanger, four air handler coils, and a jacketed kettle. The name plate loads for each piece of equipment are shown below. What size should the boiler be?

Heat exchanger: 2500 lb/h of 65 psig steam

Air handler coils: 450 lb/h of 15 psig steam

Jacketed kettle: 1850 lb/h of 90 psig steam

Answer: The total steam load is 2500 lb/h + 4 × 450 lb/h + 1850 lb/h = 6150 lb/h. The boiler should be sized to handle 6150 × 133% = 8180 lb/h which is 8180 lb/h ÷ 34.5 lb/h/boiler horse power = 237 hp. The best design might be to install two 110–120 hp boilers in lieu of one 237 hp boiler.

Selecting the right size boiler will impact the boiler room size and space allocated for the steam equipment.

Startup to steam time. The time it takes the boiler to make steam from a normal "call for steam" signal may be important. If the boiler is used in a batch application, the batch cycle time may be impacted by this consideration. Boilers with high water volumes will take longer to heat up…but will also stay warmer longer. Boiler manufactures use quick startup times as a marketing tool. This consideration will affect the boiler size and potential footprint in the boiler room.

Emissions control. This consideration is related to the boiler burner and combustion controls design. Fuel combustion systems can employ low NOx technology and oxygen trim. Low NOx systems often incorporate flue gas recycling which requires piping from the boiler stack back to the combustion air inlet. Oxygen trim will require an oxygen sensor in the stack that will need to have low voltage wiring routed back to the burner control panel. In addition, large boilers may have a CEM or continuous emissions monitoring system installed in the boiler stack. This system uses a set of sensors that provide input to a control panel to monitor boiler exhaust emissions. Many times these stack sensors are located high above the boiler and access must be available for periodic maintenance. It is a good practice to engage the environmental professionals early in a fuel-fired steam boiler design process.

Economizers. Nearly all large steam systems employ boilers with economizers. Economizers are heat exchangers located in the boiler exhaust stacking that use the hot flue gases to preheat combustion air or feed water. Installing economizers will require higher roofing in the boiler room and additional piping and ductwork. More energy-related benefits are shown in Chapter 16.

Local or remote boiler control. Until recent years, most boilers were operated locally by a dedicated boiler operator. Times have changed and today there are an increasing number of boiler rooms that have equipment operated remotely via a PLC of DCS type control system located in a control room away from the boiler room. Consequently, operators may not continuously monitor boiler room conditions. Ambient temperature, carbon monoxide, fire alarms, boiler shutdowns, pump failures, etc. may not be discovered right away. Boiler room designers need to account for these remote sensors and consider automatic lead lag systems that would autostart a redundant boiler if the primary (lead) boiler or pump fails. It is relatively easy to see that a facility manager would be able to monitor the steam generation system from their computer and receive text messages on their phone for any alarms in the boiler room.

Redundancy. Consideration should be given to redundancy. Hospital boiler systems are required to have redundancy (N+1 configuration) both in a backup fuel and total load satisfaction. In the process industry, redundancy is often desired as the consequences of losing process time generally will outweigh the cost of installing redundant steam equipment. Not all steam equipment, however, needs to be redundant. Boiler blowdown vessels and water treatment equipment is seldom installed as redundant. Boilers and feed water pumps are critical components that should be designed with backup capability. Generally it is easier and less costly to allow for redundant equipment in the design phase of a project versus having to retrofit a boiler room.

Steam quality. Steam quality was discussed in detail in a previous chapter and is another performance consideration that should not be overlooked. Clean steam is essential for certain process applications like baking, sterilizing, drug manufacture, electronic component manufacturing, and humidification. Generally, any process that the steam will come into direct contact with a product for human consumption will require steam with a dryness fraction of >99%. Steam separation or filtration may be the best way to achieve the level of steam quality desired. Applications requiring high steam quality should use a boiler with a high steam disengagement area, low steam exit velocity, and a durable enough design to allow for minimal water treatment in the boiler. Boiler room designers may have to incorporate steam accumulators into the boiler room piping design to prevent the consequences of rapid depressurization. Retrofitting a boiler room with a steam accumulation system is generally more difficult and costly than incorporating it into the initial construction.

ENVIRONMENTAL CONSIDERATIONS

Steam boilers will use fuel, radiate heat, create exhaust combustion gases, and have to occasionally discharge very hot water. The boiler room must be designed to adequately ensure all of these parameters are controlled and, designers must keep these considerations in mind while developing the boiler room floor plan.

Space concerns. Boiler room real estate is valuable and usually steam equipment competes for space in a new installation. Today's vertical boilers and compact boiler designs provide the engineer opportunities to generate a lot of steam in a very small area. One should strongly consider steam equipment maintenance when allotting space for the boilers. Large horizontal fire tube-type boilers can require several feet of clearance for tube replacement and must be taken into consideration during the room design. Overhead clearances for steam piping and stack venting are required. Since most boiler applications will require support equipment like feed water and blowdown equipment, engineers must take into considerations all the other equipment and the maintenance required on this equipment. A good practice allows at least 24″ free space around all major equipment and at least 36″ (actual NFPA code requirement) free space in front of all electrical panels. Water treatment systems need frequent adjusting and maintenance. Accessibility to these systems and room for storage/containment of chemicals will be important.

Permits. In every steam boiler system, some sort of permitting will be required. Boiler registration with a local, state, or federal agency may be required. Building permits will likely be required for new construction. All boilers will need some sort of periodic discharge of hot boiler blowdown water. This hot water will contain boiler water treatment chemicals and points of discharge need to be evaluated early in the design phase of a project. The environmental compliance folks will likely dictate where this water will eventually be sent. The hot exhaust gases need to be safely discharged to the environment. Air and water permits may need to be obtained. Fossil fuel and biomass-fired boilers may require greenhouse gas producers registration. The design engineer should always check the local codes and regulations for fuel-type usage prior to specifying a certain type of burner on a boiler.

Flue gas handling. Any boiler that combusts fuel will need to have the combustion flue gases vented to the atmosphere. The design and installation of the venting (stack) on the boiler is very important if the steam generator is to function properly. Many problems associated with poor boiler reliability can be a result of improper design or installation of the stack. Venting material selection and sizing is critical. The first step in sizing venting material is to determine the appliance rating. This rating is based on flue gas pressure and condensing potential. The four categories are shown below in Table 13.1.

It would be impractical to try and present specific venting designs here because of the vast variety of possible venting arrangements. Fortunately, there are some good stack design programs available and most stack equipment suppliers will model the boiler and boiler room to provide a good stack design. You will, however, need to obtain some basic information to correctly design the venting system:

1. Flue gas volume and temperature
2. Design ambient outside and inside temperature
3. Length of horizontal and vertical distances the vent piping must run
4. Tolerable pressure at the steam generator stack connection for all operating modes

A properly installed stack needs to provide the right draft pressure at startup and at operating temperatures. It must also prevent dangerous exhaust gases like carbon monoxide from entering the boiler room. The stack exhaust fan, if used, should be integrated with the combustion air units in the boiler room or pressure swings in the room can occur. Boiler room air pressure swings cause boiler burners to do strange

TABLE 13.1   NFPA 54 Gas-Fired Equipment Categories

Appliance Category	Vent Pressure	Condensing Potential	Temperature of Flue Gas	Common Flue Pipe Material	Annual Efficiency
I	Negative	None	>275 F	B-vent	<84%
II	Negative	Possible in the vent	<275 F	Special per manufacturer	>84%
III	Positive	Possible	>275 F	Stainless steel	<84%
IV	Positive	Possible in the heat exchanger	<275 F	Plastic	>84%

Data from National Fuel Gas Code [44].

things and will generally negatively affect boiler performance. It is highly recommended that the boilers are test fired at low, medium, and high firing periods with varying ambient air temperatures. The stack manufactures should answer all of the questions below "yes" before they supply equipment. Some general arrangement installation guidance is shown in Figure 13.4.

1. Is the common breeching sized for the maximum combined exhaust flow of all boilers being serviced by that breeching?
2. Does the breeching design maintain the correct pressure at the boilers exhaust for any combination of boiler operation?
3. Does the same pressure requirement hold true for any ambient outside air temperature?
4. Is the breeching design and installation logic tied to the boiler combustion fan?
5. Are spill switches installed at the barometric dampers that alarm or shut down a boiler?
6. Is the breeching material rated for the category of appliance assigned to the boilers?

A well-designed venting scheme may use one or more of the following; barometric damper, spill switch, exhaust fan, manual or automatic damper, and a pressure switch to help maintain a slightly negative pressure in the vent at all times even with outside temperature swings and boiler firing rates. Here are some rules of thumb that apply to venting:

1. **Stack height**: The stack height should extend through the roof high enough to avoid down drafts in the stack or the possibility of carrying combustion gases to undesirable places like air handler inlet ducts or windows.
2. **Stack size**: The stack should be sized to ensure a pressure drop does not exceed 0.2″ water column. The use of barometric dampers and/or exhaust fans may need to be incorporated into the design to ensure adequate pressure drop during a wide range of firing rates and outside air temperatures.
3. **Breeching connections**: The boiler breeching connection to the stack should be made whenever possible with a 45° upward sweeping elbow.
4. **Multiple boiler stack**: When two or more boilers are connected to a common stack, they should never enter the stack at the same height and they should be isolated from each other. A common method to achieve separation is by the use of sequencing draft controls.
5. **Breeching**: Breeching should be as short as possible, made from round duct work whenever possible, and have a slight elevation to the stack connection.

Stack sizing should be specific for each boiler application. Table 13.2 includes some general sizing guidelines for single and multiple boiler applications.

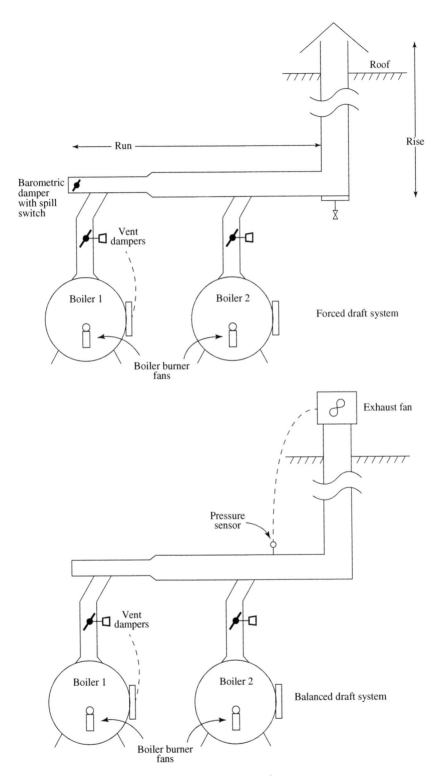

FIGURE 13.4 Steam boiler venting system general arrangement.

TABLE 13.2   **Boiler Breeching and Stack Sizing Guidelines**

Minimum Diameter Stack	Diameter for 100 ft Run			
	Number of Boilers			
Individual Boiler HP	1	2	3	4
15–20	6″	9″	10″	11″
15–40	8″	11″	13″	14″
50–60	10″	13″	15″	17″
70–100	12″	16″	19″	21″
125–200	16″	21″	24″	28″
250–350	20″	26″	32″	34″
400–600	24″	32″	38″	42″
700–800	24″	38″	44″	48″

Data from Cleaver-Brooks Inc. [20].

Hot water discharges. The boiler will have a blowdown rate that will be determined based on the effective evaporation rate and the boiler manufactures recommended solids level. This hot blowdown water is generally considered a waste stream and has to be disposed of in accordance with the local environmental regulations. Typically the drains that will receive this water should not be constructed of PVC or any other material that is affected by water temperature up to 140 F. In addition we know that the boilers will need to be drained periodically for maintenance and inspection. **It is a good rule of thumb to size the drains to be able to handle at least 2 to 3 times the water usage rate in the boiler room**. Water usage rate should include drain, backwash and purge rinse rates of all equipment. Blow off separators should be equipped with automatic cooling to prevent water dischargers greater than 140 F.

Boiler room temperature control. The boiler room temperature is very important due to the electrical equipment longevity and personnel comfort. A hot boiler room (>120)F will shorten the life of electrical components and be unsafe for any prolonged human activity. Consequently, adequate ventilation with outside air or chilled air is usually required. If combustion air for the boiler uses room air, then air louvers to the outside or powered combustion air units must be sized accordingly. The boiler room must never become negative or dangerous flue gases could spill into the room and create a harmful environment. When makeup air handlers are employed to keep the boiler room cool, then their operation should be logic tied to the boiler burner's fan, otherwise room pressure could impact the burner's performance.

HVAC engineers will need to size boiler room ventilation units to counterbalance the radiation heat losses from the boiler room equipment. Engineers should determine the square footage of exposed hot metal in the room, determine the average delta temperature between the hot metal and ambient, and then a **good rule of thumb is 3 btu/h/ft²-ΔF to calculate heat addition to the room**. The ΔT is the difference in temperature between the hot metal and the room temperature. Equipment manufacturers may be able to provide radiation loss data for their equipment.

Example 13.2

A maintenance manager is concerned about the HVAC cooling load in the boiler room and has asked you to find out how much heat is being radiated by the steam system. You use an infrared heat gun to determine the surface temperature of the various equipment in the boiler room. In addition, you estimate the areas that are hot and obtain the following data. The boiler room needs to be keep at 85 F.

Equipment	Surface Area	Surface Temperature	ΔT	Heat Radiated (@3 btu/h/ft²-F)
Boiler shell	630 ft²	125 F	40 F	75,600 btu/h
Boiler piping	15 ft²	330 F	245 F	11,025 btu/h
Deaerator	320 ft²	100 F	15 F	14,400 btu/h
Feed water lines	76 ft²	208 F	123 F	28,044 btu/h
Blow off vessel	18 ft²	125 F	40 F	2160 btu/h
Steam piping	84 ft²	330 F	245 F	61,740 btu/h
Total heat load				192,969 btu/h

Answer: You better have a cooling system that can remove 192 K btu/h. This calculation is based on moderate air movement around the hot surfaces. High air movement in the room will promote convection heat transfer around the hot equipment may cause additional heat losses.

This heat can be removed via active cooling (i.e., chilled water or air conditioning) or by outside air ventilation. A simple method of removing heat from a space is passive ventilation using outside air. The accepted method of calculating the amount of outside air is to the general ventilation formula: **CFM = heat load (btu/h)/1.1 × ΔT**.

CFM is the outside air flow into the room, and ΔT is the difference in temperature (F) between the outside air and the desired room temperature.

Example 13.3

Calculate the amount of outside air needed to ventilate the room mentioned in Example 13.2.

Answer: With 192,969 btu/h head load, and 95 F worst case outside air temperature, you would need 192,969 btu/h/1.1 × 95 – 85 F = 17,542 CFM air flow for those conditions. Obviously, during cooler outside air temperatures, a much lower ventilation rate is required. The use of VFD fan motors is warranted.

Steam and gas line venting. The last environmental consideration is design criteria to address steam and fuel line venting. Gaseous fuel trains will have regulators, and the boiler safety relief valves (SRV) will need to be vented outside. The termination points should not be above or near where people can walk. **If local codes allow,**

multiple vent lines can be connected to a common line, provided the common vent line cross-sectional area is equal or greater than the sum of the individual lines feeding into the common line. Generally, codes do not allow a valve to be installed between the vent connection on the component and the environment.

Example 13.4
The boiler room being designed will house two natural gas-fired boilers each having a gas regulator with a ½ inch vent connection. The closest outside wall is over 200 feet away and the installing contractor wishes to run one common gas vent line. How big does the line have to be?

Answer: Anytime a common line is used to vent multiple components, then the cross-sectional area of the common line must be equal or greater than the sum of the cross sections of the individual lines. Therefore the common line must have a cross-sectional area equal to 2 Πr² or 2 × Π × (0.25 inch)² = 0.3925 inches or 0.40 inches. A ¾ inch line will have a cross-sectional area of 0.44 inches, therefore is large enough.

In some cases smaller gas regulators can be configured as ventless by installing a vent limiter device on the regulator. I recommend you consult with the local code official and insurance supplier before this method of regulator venting is designed.

BOILER ROOM UTILITIES

The boiler room must be designed with the following utility considerations.

1. Enough combustion air supply
2. Right quality and quantity of makeup water
3. Right drainage
4. Adequate fuel supply and pressure for the boiler(s)
5. Adequate electrical power to operate the boiler room equipment

Combustion air requirements. Each boiler that uses fuel will require air to be mixed with the fuel to support combustion. Each type of fuel will require a minimum amount of air. This combustion air must come from within the boiler room or come from the outside (sealed combustion). The boiler room designer must ensure adequate combustion air quantity and strive for the highest combustion air temperature. Combustion air and flue gas volumes for some common fuel gases, solids and fluids are indicated in Table 13.3.

Unfortunately, the minimum amount of combustion air as shown in Table 13.3 is theoretical and in the real world will not ensure complete combustion. Therefore, we must add some excess air to achieve complete combustion. The relationship between combustion air and flue gas chemistry is shown below in Figure 13.5. The

TABLE 13.3   **Comparative Data for Various Fuels**

Fuel	Heating Value	Theoretical Air for Combustion	Theoretical Flue Gas Volume	Ultimate CO_2 Concentration in Dry Flue Gas
	btu/lb	ft³/lb	ft³/lb	
	(btu/gal)	(scf/gal)	(scf/gal)	
	[btu/scf]	[scf air/scf fuel]	[scf/scf gas]	%
Anthracite coal	12,680	150	152	19.9
Wood	6,300	70	80	20.3
Butane	[3225]	[30.5]	[32.9]	14.3
Natural gas	[1000]	[9.5]	[10.5]	11.7
Digester gas	[690]	[6.4]	[7.8]	14.5
Propane	(91,500)	(851)	(934)	13.8
Methanol	(64,630)	(559)	(681)	15.0
Gasoline	(123,361)	(1183)	(1272)	15.0
#2 oil	(137,080)	(1354)	(1440)	15.7
#6 oil	(153,120)	(1484)	(1558)	16.7

Data from *North American Combustion Handbook* and from Eclipse Inc. [11, 35].

ideal range for excess air is usually around 15% (or 115% of total required theoretical air). By measuring the amount of CO, CO_2, or O_2 in the flue gas, we can determine the degree of combustion and make adjustments accordingly. **High levels of flue gas CO indicate the burner is operating lean**. High levels of flue gas O_2 indicate high excess air. Adding too much excess air will have a negative impact on combustion efficiency. Heating the extra air up to the flame temperature by the combusted fuel will rob energy that could be used for heat transfer to the boiler water. **Generally a good rule of thumb is a decrease in 10% excess air will improve combustion efficiency by 1%**. Operators should adjust burners to yield low amounts of CO and O_2 in the flue gas.

Consequently, combustion air preheating will also impact efficiency. When the combustion air is preheated less energy is used to heat that air up to the combustion gas temperature. **The rule of thumb for preheating combustion air is a 1% combustion efficiency increase for every 40 F temperature rise in the combustion air**. Engineers should design and operators should strive to keep combustion air as warm as possible. Using stack economizers or pulling hot air from the boiler room ceiling are a couple methods to preheat combustion air.

If boiler room air will be used as combustion air there will need to be fresh air opening in the walls to prevent the room from becoming negative pressure. A general arrangement is shown in Figure 13.6. **The rule of thumb for fresh air openings is 1 sq. inch opening per 4000 btu fuel input, with at least two openings per room** [45]. Another method of calculating fresh air openings is to use area required in sq. feet = CFM/FPM, where CFM is cubic feet per minute (volume) and FPM is feet per minute (velocity). For combustion air use total BHP × 16 CFM/BHP and for

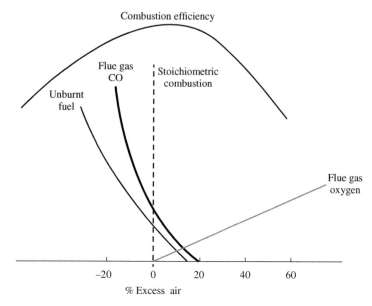

FIGURE 13.5   Relationship between combustion air, fuel, and efficiency.

ventilation air use total BHP × 4 CFM/BHP or a total of 20 CFM/BHP. Acceptable air velocities are 250 FPM for low openings and 500 FPM for high openings. The opening louvers dampers should be logic tied to the boiler combustion fan operation. All outside wall openings should also be covered with a 1/4 inch or larger opening screen to keep out large insets and birds.

Example 13.5

Calculate the fresh air opening requirements for a new boiler room housing three 60 bhp steam boilers. Each boiler has a fuel input of 2680 MMbtu and will take its supply of air from the boiler room.

Answer: Using the 1 sq. in/4000 btu input you would need 2,680,000 × 4/4000 = 2010 sq in of opening. This equates to two openings each having a free space of 7 sq. ft. Also, we can determine the opening size by using the CFM/FPM method. Here we can use 20 CFM/BHP × 180 BHP = 3600 CFM. If we use 250 FPM as the acceptable velocity, then our opening would be calculated at 3600/250 = 14.4 total sq. ft or 7.2 sq. ft/opening.

Note: The calculated opening size refers to free space in the opening which may represent only about 60–75% of the actual opening size. You should always consult with the louver manufacture to determine actual free space.

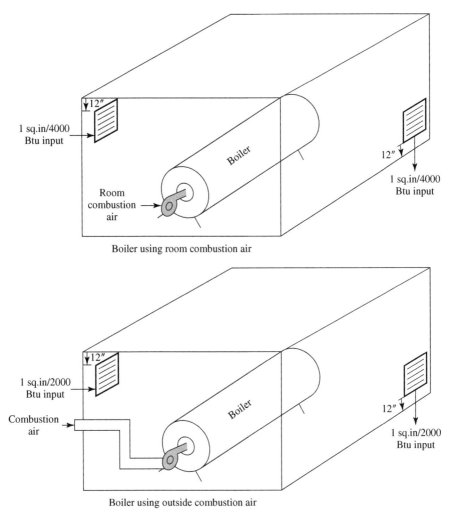

1 sq.in/4000
Btu input

Room
combustion
air

Boiler

12″

1 sq.in/4000
Btu input

Boiler using room combustion air

1 sq.in/2000
Btu input

Combustion
air

Boiler

12″

1 sq.in/2000
Btu input

Boiler using outside combustion air

FIGURE 13.6   Fresh air opening designs for a boiler room.

One ventilation question that confuses engineers is how do we account for both room ventilation and combustion air ventilation. It depends on where the source of combustion air comes from. If you develop an air balance in the boiler room it might clear things up. When boilers use sealed combustion, no combustion air supply in the room is required and room ventilation is calculated as per the general ventilation formula shown previously. When, the boilers take combustion air from the room, the combustion air CFM can be subtracted from room ventilation CFM.

Water usage. Makeup water will be used by the boilers and must be supplied to the boiler room. The amount of water needed in the boiler room to service the steam system will be the sum of the boiler fresh makeup water, the blowdown water, and blowdown cooling water volume. All steam boilers use water and will need a supply of clean makeup water. The boiler fresh makeup water volume will be the evaporation rate of the steam boilers minus the amount of condensate returns. We know that each boiler will evaporate up to 34 lb of water/bhp to make steam. **The total boiler hp × 34 lb/bhp minus any condensate return volume will be the effective evaporation rate of the steam boilers**. If we have 50% condensate returns, then the effective evaporation rate is only 34 lb/hp × 50% = 17 lb/bhp or 17 lb/8.34 lb/gal = 2.04 gals/bhp.

The blowdown volume will largely depend on the manufactures recommended solids level in the boiler water, the amount of solids in the fresh makeup water, and the percentage of fresh makeup water. In addition, the blowdown volume will need to be tailored to the chemical treatment system selected. Consequently, the design engineer needs to only estimate blowdown volume. The estimation of blowdown volume can be calculated using the formula:

$$BD = ER / COC - 1,$$

where

BD is the blowdown rate in lb/h
ER is the effective evaporation rate in lb/h
COC is the desired **cycles of concentration**

The COC is a ratio of solids in the boiler water to the solids in the fresh makeup water. If the boiler manufacture states the solids level should not exceed 2500 ppm and the fresh water makeup has 250 ppm, the desired COC will be 2500/250 or 10 cycles.

The blowdown cooling water volume will need to be calculated and included in the total water usage. The hot boiler water being blowdown is usually drained to a flash vessel or blow off separator, cooled to less than 140 F and discharge to the drain. The flash vessel will depressurize the hot boiler water to atmosphere pressure which will lower the boiler water to 212 F. The cooling of this water to 140 F can be calculated by knowing the maximum blowdown rate in lb/h and cooling water temperature according to the formula:

$$TW = \left(BD \times T_{bd}\right) + \left(CW \times T_{cw}\right) / T_{tw}$$

where

BD = blowdown rate in lb/h
T_{bd} = the hot boiler water temperature (212 F)
CW = the cooling water flow in lb/h
T_{cw} = cooling water temperature
T_{tw} = final temperature of the total water drained (140 F)
TW = the total water drain rate in lb/h

Example 13.6
You have to determine the total water usage of a steam system for a new boiler room that will house a 400 hp steam boiler. The anticipated condensate returns will be 25% (75% makeup rate) and the fresh water make up has a temperature of 65 F and

a solids level of 250 ppm. The boiler manufacture indicates a total dissolved solids level of 3000 ppm or less is to be maintained in the boiler for warranty. The boiler blowdown water will be sent to a flash tank that will use fresh city water as the cooling water. What will be the total water usage rate?

*Answer: The total maximum water usage rate will be the sum of the effective evaporation rate of the boiler, plus the boiler blowdown rate, plus the blowdown cooling water rate. The effective evaporation rate can be calculated as ER = 34 lb/h × 400 hp × 0.75% = **10,200 lb/h** steam flow, and the blowdown rate can be calculated by BD = ER/COC − 1 where the COC is 3000/250 or 12 cycles. So BD = 10,200 lb/h/1 − 12 = **927 lb/h**. Lastly, the cooling water flow can be calculated using TW = (BD × T_{bd}) + (CW × T_{cw})/T_{tw}. We know BD = 10,200 lb/h, T_{bd} will be 212 F. Temperature of the makeup cooling water is 65 F, and the desired drain water (TW$_t$)temperature will be 140 F. We also know that the cooling water flow rate = the total water drain rate − the BD flow rate. Substituting in CW = TW − BD, we then have TW = BD (T_{bd} × T_{cw})/(T_{tw} − T_{cw}) and when we substitute to known values, we have, TW = 927 lb/h × (212 F − 65)/(140 − 65) = 1817 lb/h./. The cooling water flow rate is the TW − BD rate which will be 1817 lb/h − 927 lb/h = **890 lb/h**.*

*The water line into the boiler room must be able to handle at least **10,200 lb/h + 927 lb/h + 890 lb/h = 12,017 lb/h or 24 gpm**.*

Fuel supply. The steam generators in the boiler room will likely require a gaseous, liquid, or solid fuel to feed the burner. Gaseous and liquid fuels will need to be delivered to the boiler(s) at the right pressure and quantity where solid fuel will need to be delivered at a constant flow with some level of consistency. Burners that use natural gas or propane will require a minimum inlet pressure. Designers need to make sure gas mains and regulators are sized (see Chapter 8) to provide adequate gas flow, especially when multiple gas-using equipment is connected to a common main. When oil burners are used, the oil supply and return system needs to match the manufactures specifications for an oil delivery system. If an oil day tank is used, then the level control system of the day tank must be logic tied to the larger oil storage tank. Appendix A has gas and oil line capacity charts that can be used to size mains. A typical oil delivery setup is shown in Chapter 8. Likewise, a biomass boiler will require constant feed of wood chips, refuse, or other fuel. Where the solid fuel is stored and how it is delivered to the boiler will influence where the boiler (s) is located. You will need to have access to the fuel you wish to use and the boilers will need to be installed in an area that will accept the use of this fuel. The use of stored gases like liquid propane is sometimes forbidden in some parts of the USA. This is due to the heavy (relative to air) nature of LP gas and the risk of a gas leak, build up, and explosion. Local codes for fuel usage and storage should be consulted early in the design process.

Electrical power. The boiler room will use a lot of electrical power. Burner fans, feed water pumps, combustion air units, control panels are a few of the electrical loads in this room. Designers need to account for the three phase, single phase, and the low voltage control power requirements. It is a good practice to overdesign and allow for future expansion. Two frequently overlooked electrical design features are E-stop switches and electrical disconnects. Some local codes require Emergency (E) stop switches be placed near all major powered equipment. Local electrical disconnects (lock out tag out boundary) should always be placed in easy to access areas near each fan motor or pump, etc. Also, provisions for electrical communication with process computers or building management systems should be considered up front in the design phase. Electrical design specifications should require separate high and low voltage conduit runs. For electric boiler installations, ensure ample power is available at the right voltage and frequency (i.e., 60 vs. 50 hz) to operate the boiler and support equipment.

14

STEAM SYSTEM APPLICATIONS

We have discussed the usefulness of steam as a heat transfer medium and described the subsystems in detail in previous chapters. Collectively, that information is very useful only if it can all be applied to make a complete steam system. This chapter will review the specific type of steam systems and some of the design considerations required for each type. Some specific applications are shown below in Table 14.1. It is relatively easy to see that process steam systems are used to make products we use every day. Although there are dozens of steam system configurations in use today, generally steam system configurations can be described in one of these four type of applications.

- Low-pressure steam with high condensate returns
- High-pressure steam with high condensate returns
- High- or low-pressure steam with little or no condensate returns
- Very high-pressure or superheated steam with some condensate returns

It is not uncommon to see a combination of applications from different types employed by a single steam system (i.e. high-pressure process steam used in conjunction with low-pressure steam heating coils). Each of these types of steam application requires a unique set of design criteria for optimal performance. In addition to the four applications, considerations for interconnecting multiple boilers of any pressure rating with a common condensate/feed water system are described. Lastly, specialized steam-using equipment like a back pressure turbines, hydro heaters, reboilers, sterilizers, etc. are described and guidance for their installation is presented.

Process Steam Systems: A Practical Guide for Operators, Maintainers, Designers, and Educators, Second Edition. Carey Merritt.
© 2023 John Wiley & Sons, Inc. Published 2023 by John Wiley & Sons, Inc.

TABLE 14.1 Steam System Applications

Equipment	Process Application	Industry
Condenser	Steam turbine operation	Aluminum, chemical manufacturing, forest products, glass, metal casting, petroleum refining, and steel
Distillation tower	Distillation, fractionation	Chemical manufacturing, petroleum refining, biorefinery
Dryer	Grains/solids and slurry drying	Forest products/biorefinery
Evaporator	Evaporation/concentration	Chemical manufacturing, forest products, petroleum refining
Process heat exchanger	Process air and liquid heating, light ends distillation, storage tank heating. Hot water generation, water for injection (WFI)	Aluminum, chemical manufacturing, forest products, glass, metal casting, petroleum refining, steel, food, and pharmaceutical preparation
Reboiler	Fractionation	Petroleum refining, biorefinery
Reformer	Hydrogen generation	Chemical manufacturing, petroleum refining
Separator	Component separation	Chemical manufacturing, forest products, petroleum refining
Steam ejector	Condenser operation, vacuum distillation	Aluminum, chemical manufacturing, forest products, glass, metal casting, petroleum refining, and steel refining
Steam injector	Agitation/blending, heating	Chemical manufacturing, forest products, petroleum refining, food preparation
Steam turbine	Power generation, compressor mechanical drive, hydrocracking, naphtha reforming, pump mechanical drive, feed pump mechanical drive	Aluminum, chemical manufacturing, forest products, glass, metal casting, petroleum refining, and steel
Stripper	Distillation (crude and vacuum units), catalytic cracking, asphalt processing, catalytic reforming, component removal, component separation, fractionation, hydrogen treatment, oil processing	Chemical manufacturing, petroleum refining
Thermo-compressor	Drying, steam pressure amplification	Forest products
Steam press/iron	Laundry cleaning	Hotel/hospital/uniform laundries
Steam tunnels	Shrink wrap plastics, laundry washing	Packaging and commercial laundries
Steam snuffing	Fire protection	Chemical processing and refining
Thermoforming	Plastics molding	Plastic parts manufacturing

Data from the Advanced Manufacturing Office, US Department of Energy [25].

LOW-PRESSURE STEAM WITH HIGH CONDENSATE RETURNS

These types of systems are used widely in the steam heating, steam to hot water generation, or heat tracing applications. Typically the steam used is less than 30 psig with 10–12 psig found most often. Low-pressure steam is generated and sent throughout a building or series of buildings where the steam is used for heating. Schools, colleges, and older manufacturing facilities have used low-pressure steam heating systems for many years. These types of systems share the same characteristics of a high-pressure system with high condensate returns, only the steam quality can be much lower. It is not uncommon to see a high-pressure process steam system used with pressure reducing valves to lower to steam pressure for space heating in the same facility. Many low-pressure steam heating systems are being replaced with high efficient hot water systems. When a process steam application calls for a process temperature <230 F, low-pressure steam can be good heat transfer method. Because of the high returns, water chemistry is relatively easy to maintain. In some areas of the United States, low-pressure steam systems do not require a stationary operator, not the case for some high-pressure steam systems. A typical low pressure steam system is shown in Figure 14.1

Example 14.1

A new addition to college dormitory is being conceptually designed and the existing campus low-pressure steam system will be used to heat the dormitory. You are being asked to size the steam load. The total area to be heated will be about 5000 square feet. While a detailed analysis of room heat losses and air changes would need to be

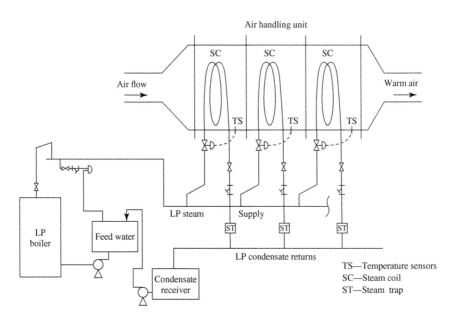

FIGURE 14.1 Low-pressure steam systems with high returns.

developed to make an exact sizing calculation, some generalizations can help estimate the steam load for this application.

Answer: Because the dormitory will have a relatively high window area and three of the four walls are outside walls, you assume good insulation but only average air tightness. If we use the standard of 25–40 btu per hour per square foot [46] for heating, then we would need 5000 × 40 = 200,000 btu/h. If we add 30% safety factor, then 260,000 btu are required. Since 1 lb of steam will provide about 1000 btu, then we will need 260 lb of steam per hour or 7.5 bhp.

One caution to keep in mind for these low-pressure steam system is condensate lift limitations. If condensate is required to be pumped any significant distance or height, the low-pressure steam may not have enough motive force to push the condensate. If the steam pressure at a steam trap is only 10 psig, then a rise of more than 10–12 feet is enough to create water logging. In these applications, you must employ a hartford loop system or drain the condensate to an atmospheric tank and electrically pump the condensate back to the boiler room. Aside from the condensate lift restriction, low-pressure steam systems can suffer from potential steam quality issues. Boiler water carryover will occur if steam demand is not regulated or boiler water chemistry (i.e. high alkalinity) gets out of control.

Low-pressure steam systems with high returns can also employ a "**hartford loop**" design. This essentially is a closed loop system that gravity drains the condensate back to the boiler. Figure 6.2 shows a typical Hartford loop design. Many older steam heating systems use this system configuration. A hartford loop system is simple and relatively maintenance free. Steam heating used in multistory buildings is a good fit for a hartford loop design.

HIGH-PRESSURE STEAM WITH HIGH CONDENSATE RETURNS

This type of steam application is considered a "closed" system and is the most widely used of steam in the process industry. An example of a closed high-pressure steam system is shown in Figure 14.2. High-pressure steam is usually steam at or under 300 psig. Steam in this application can be used for product heating via a shell and tube heat exchanger, steam coil in a tank, plate a frame heat exchanger, jacketed kettle, steam drum/roller, evaporator reboiler, steam jacket, steam press, or steam ironer. In most cases the steam condenses in the equipment and the condensate is removed, recovered and returned to the boiler room. Applications will have the following characteristics:

1. Nearly 100% condensate return
2. Minimal freshwater makeup
3. Minimal boiler blowdown volume
4. Require relatively high steam quality
5. Require minimal water treatment

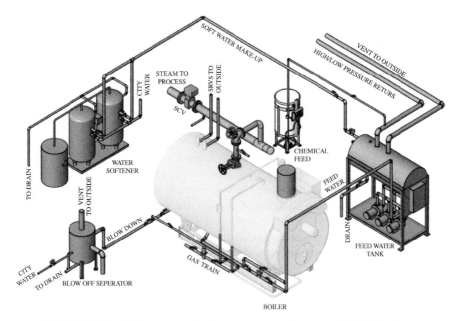

FIGURE 14.2 High-pressure closed steam system. Courtesy of C&S Engineers.

Example 14.2

A batch process used to plate metal parts requires a tank of plating solution to be heated up to 285 F using a steam coil. The 2010 gallons of plating solution has a density of 9.35 lb per gallon and is 125 F prior to initiating the heating process. The plating cycle is only 15 min and occurs once every 24 h, but the heat up period from the 125 to 285 F is 1 h. The plating solution has a specific heat 0.85 btu/lb-F. How much steam at 100 psig pressure do we need?

Answer: First, determine the amount of energy needed to bring the plating solution up to temperature in one hour. Since no phase change will occur, only sensible heat will be added. The 2010 gallons will weigh 2010 gallons × 9.35 lb/gal = 18,794 lb. This 18,794 lb of solution has a 0.85 btu/lb-F specific heat. Therefore, to heat this solution up to 285 F from 125 F in one hour will require.

$$18{,}794 \; lb \, / \, h \times 0.85 btu \, / \, lb - F \times \left(285F - 125F\right) = 2{,}555{,}984 \; btu \, / \, h$$

Since the plating solution must reach 285 F, we must have a steam heating source hotter than 285 F. Steam at 100 psig will have a saturation temperature of 328 F and a latent heat of 889 btu/lb. The amount of 100 psig steam would be 2,555,984 btu/h/889 btu/lb steam heat = 2875 lb/h. A good design would add at least 30% extra capacity for a total load of 2875 lb/h × 1.30 = 3738 lb/h steam.

Note

1. The aforementioned Example 14.2 does not account for the heat necessary to heat the tank metal from 125 to 285 F, nor does it account for any radiant heat losses during the heat up cycle. Both of these would need to be considered (some of the reason for the 30% extra capacity).
2. If the heat up time is required to be 30 min vs. 1 h, then the steam load would double to 3450 × 2 = 6900 lb/h (equals 3450 lb/30 min).

These high-pressure high condensate process applications require good steam trap sizing and conservative condensate pipe sizing; otherwise, water logging can occur in the steam-using equipment. Once the system is set up to operate properly, they usually are quite easy to maintain because the steam and condensate flow is relatively steady.

The challenge with these systems is the improper sizing of the boiler. Batch applications can lead to excessive boiler burner cycling, especially if the burner is oversized or cannot reach a low turndown rate.

HIGH- OR LOW-PRESSURE STEAM WITH LITTLE OR NO RETURNS

These types of steam systems, similar to the one shown in Figure 14.3, are found in sterilization, food processing, biofuels, paper manufacturing, and any other application where steam is injected directly into the product being heated or the condensate must be discarded. Some applications, like using steam to heat plating tanks of acid or steam tunnels for laundry, do not recover the condensate because of the risk of the condensate contamination. Steam systems with little or no returns will

1. Have very high fresh water makeup rate
2. Have high boiler blowdown rate
3. Require more water treatment technology
4. Require extremely high-quality steam

This application requires some distinct design considerations to protect the product being heated. For example, sterilizers that use direct injection steam to disinfect cannot tolerate wet or dirty steam. Imagine a doctor picking up a surgery tool that has been sterilized this way that has dried dissolved solids spots on it. Therefore, any water and contaminants in the steam needs to be removed. Steam filters and separators are commonly employed in these applications. Boiler sizing also become a bit tricky. Since there is a high percent of colder makeup water, more fuel energy will be used to heat the water up to the boiling point before any steam is produced. This added sensible heat must be taken into account or the boiler can be designed undersized. In many applications of this type, the steam is used in batch-type applications and the use of steam accumulators is warranted. Such is the case for autoclaves that

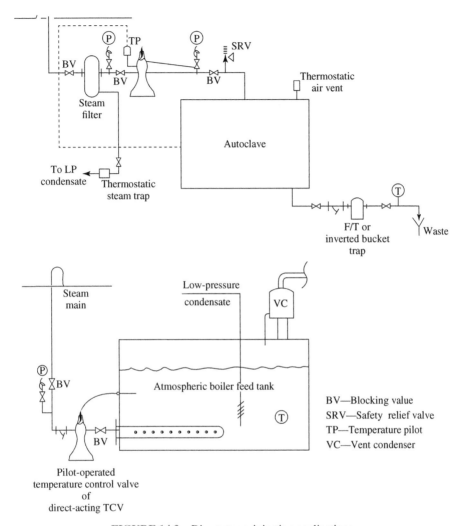

FIGURE 14.3 Direct steam injection applications.

use steam to sterilize. The autoclave vessel initially is cold and will have a warm up load that could exceed the boiler capacity.

Example 14.3

An application for a steam sterilizer requires 4250 lb of steam per hour. Since all the steam is direct injected and the condensate contaminated, there will not be any condensate returned. The fresh softened water makeup is 70 F and the boiler is a low-pressure 15 psig design. How big does the boiler have to be? Our application will use 70 F makeup water and operate at 13 psig.

Answer: Boiler manufactures rate their boilers based on 212 F makeup water and 0 psig
operation. First we must calculate how much energy will be required to make 1 lb of 13
psig steam from 70 F water. From the steam charts, steam at 13 psig will have a temper-
ature of 246 F and a total energy content of 1162 btu/lb. Therefore, 1 lb of 13 psig steam
will have an enthalpy = 1162 btu. We know 1 lb of water at 70 F will contain 38 btu. So,
to make one lb of 13 psig steam from 1 lb 70 F water will require 1162 – 38 = 1124 btu.
If we look at a boilers output in btu (taken from a manufactures data sheet), then divide
the output by 1124 btu/lb, we can determine the actual lb of steam from a boiler operating
under these conditions. This application will require a boiler with 4250 lb/h × 1124 btu/
lb = 4,777,000 btu output.

In addition, some enthalpy will be lost in the hot blowdown water. Using the rule of
thumb for 0% returns, the boiler will have a blowdown rate approximately 72
gals/h/100 hp. Since 34,500 btu = 1 bhp, the boiler size is 4,777,000 btu/34,500 btu/
bhp = 138 bhp. The blowdown rate is then calculated as 72 gals/h/100 bhp × 138
bhp = 99 gal/h. The saturated water/steam table in Appendix A show that blowdown
water at 13 psig will have an enthalpy of 214 btu/lb. The heat loss from blowdowns
is 214 btu/lb × 100 gal × 8.3 lb/gal = 177,620 btu/h. The boiler for this application
*must have a minimum output of 4,777,000 btu/h + 177,620 btu/h = **4,954,620 btu/h**.*

It should be quite apparent that a steam system application that has little or no
condensate returns requires the boiler size calculation take into account the energy
required to heat up the cold water makeup and account for the heat loss from the
blowdown water. In the application above, if the boiler was sized solely in manufactures data, then we would have calculated the boiler to be 4250 lb/h × 1/34.5 lb/h/
bhp = 123 bhp. Considering the high relatively cold makeup water and the high blowdown rate, the actual boiler size needed is close to 4,966,770 btu/34,500 btu/
bhp = 144 bhp. Alternatively, if we had applied the 130% rule, using the manufactures output data, we would have specified a 123 × 130% = 160 bhp.

HIGH-PRESSURE OR SUPERHEATED STEAM WITH CONDENSATE RETURNS

This type of system is found in the process industry where the process requires relatively
high temperatures. Rubber/plastics processing, refineries, chemical manufacturing,
and other process industries use these types of systems. The high steam pressure may
require expensive steam system components, and superheated steam use will require
careful condensate collection systems. Remember that superheated steam is saturated
steam with added sensible heat energy. Therefore superheated steam must give up
some sensible and latent heat to fully condense. Correct trap selection is critical. These
types of steam systems can be susceptible to steam hammer and steam erosion across
steam regulating valves. Equally important is to design for thermal expansion if long
pipe runs are required. It is imperative that drain legs be installed at regular intervals to

support condensate draining prior to startup. Individual condensate runs should also be introduced into the main condensate header via a condensate diffuser to minimize steam hammer. Applications for using high-pressure or superheated steam are shown in Figure 14.4

MULTIPLE BOILER INSTALLATIONS

Regardless of the amount of condensate return volume, when a design calls for multiple boilers some additional design features need to be considered. A boiler master control panel as described in Chapter 8 should be included to ensure lead lag sequencing of the boilers occurs. In addition, system operators should strive to operate any steam boiler in its most efficient output range which is usually 20–80% output. When multiple boilers are being tasked, the boiler master control panel will help achieve this goal. When multiple boilers are connected to a common steam header, high water protection is required for each boiler. The reason for this is migrant steam from an operating boiler will flood an idle boiler unless a check valve is installed in the boiler steam line or, a means to keep equal water levels in all boilers is used. An equalization line between the operating boilers can be used to prevent idle boiler flooding as shown in Figure 14.5.

When a single feed water pump is used to supply water to multiple boilers a means to prevent inadvertent filling of the idle boiler must also be considered. Control valves with minimum recirculation lines installed are the best means to accomplish this. Figure 14.6 shows a recommended piping schematic.

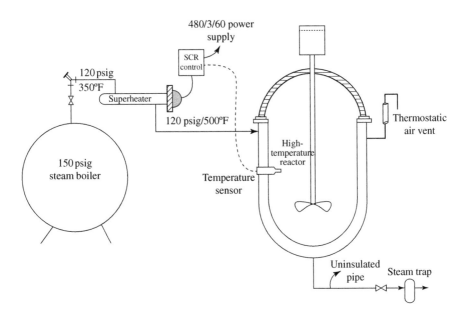

FIGURE 14.4   System using high-pressure or superheated steam.

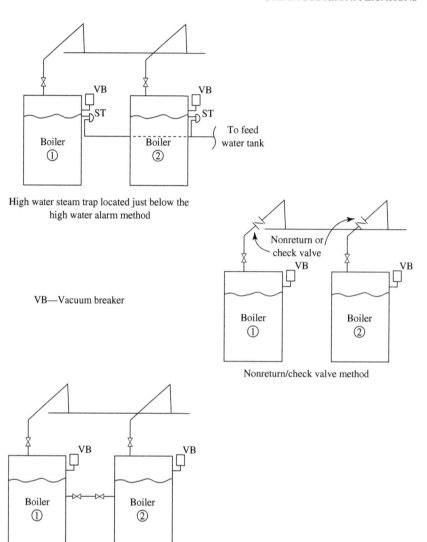

FIGURE 14.5 Multiple boiler high water protection design.

There may be occasions where multiple boilers at different locations are being fed from a different feed water tanks but have common condensate collection. For instance, a municipal trash burning plant is located near a manufacturing plant. Steam come from both plants share a common steam distribution system. Here, the condensate needs to be divided to supply both plants. One way to accomplish this is to use an overflow condensate receiver design. Whichever steam system is considered

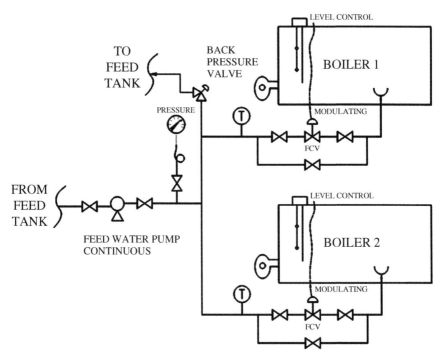

FIGURE 14.6   Multiple boilers fed from single feed water pump. Courtesy of C&S Engineers.

primary (generally the manufacturing plant) will get all the condensate required to support their feed water system. The other boiler system will get the overflow condensate. Figure 14.7 shows a suggested piping scheme for such a design.

Steam for Hot Water Generation

When a plant has copies amounts of steam available and a need for hot water, a steam to hot water generator system can be a good application. Industries that use a high volume of hot water for CIP (clean-in-place) or general wash down water in high volume short duration batches are good applications for this type of hot water generation. Some domestic hot water systems use this setup. This setup is really useful if there are periods of low steam use and the steam is being generated via a big boiler. In lieu of the boiler cycling profusely, the hot water generator can provide a steam load to reduce the cycling. The system is quite simple. A steam supplies heat to a heat exchanger that will heat up a hot water loop. In addition to the heat exchanger, hot water pumps, hot water storage system, and a makeup water control system is required for the hot water loop. The size of the heat exchanger and storage tank will depend on the hot water usage volume and rate. A typical system set up is shown below in Figure 14.9.

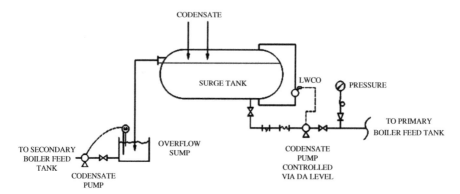

FIGURE 14.7   Common condensate collection for multiple feed water systems. Courtesy of C&S Engineers.

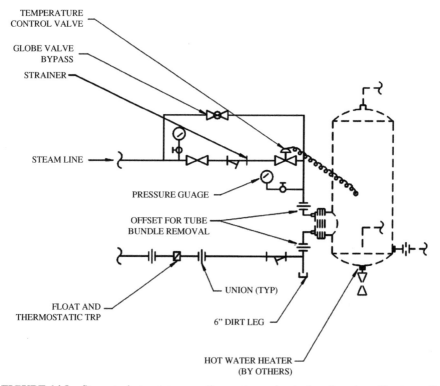

FIGURE 14.8   Steam to hot water generation system using tank coil system. Courtesy of C&S Engineering.

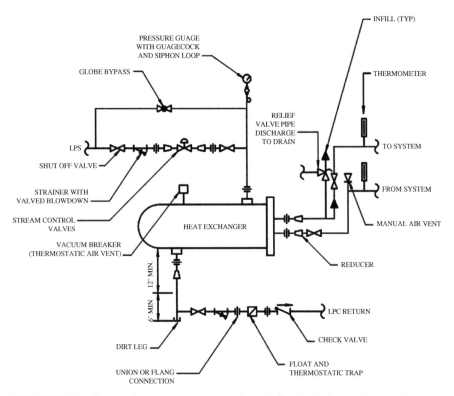

FIGURE 14.9 Steam to hot water generation using a shell and tube heat exchanger. Courtesy of C&S Engineering.

Other Miscellaneous Steam-Use Application Designs

Steam can be used to heat up air in an air handler via a steam coil. Figure 14.10 shows a typical design detail. Steam is also used frequently to heat up a liquid product via a shell and tube heat exchanger as shown in Figure 14.11. Steam is generally used on the tube side while the fluid being heated is on the shell side.

Steam is sometimes used to provide a means to humidify air. A typical steam humidification system is shown below in Figure 14.12.

Steam is often used to heat warehouses and loading docks. Such applications will use a ceiling mounted unit heater. A typical design detail is shown in Figure 14.13.

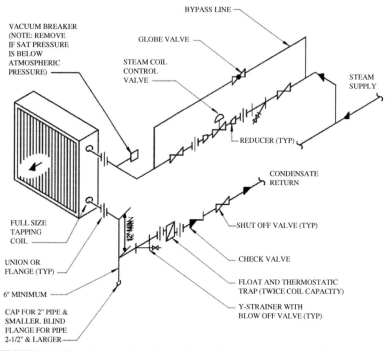

FIGURE 14.10 Steam coil for air handler piping detail. Courtesy of C&S Engineering.

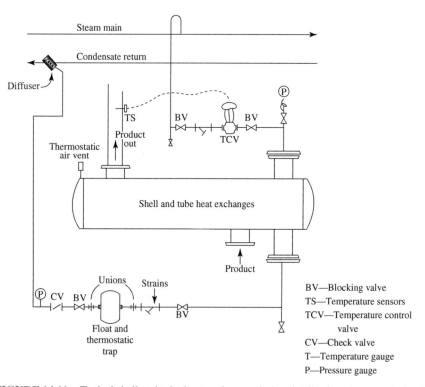

FIGURE 14.11 Typical shell and tube heat exchanger design detail using steam as the heat source for process heating.

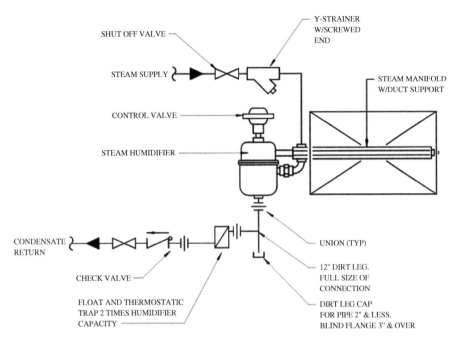

FIGURE 14.12   Steam humification piping detail. Courtesy of C&S Engineers.

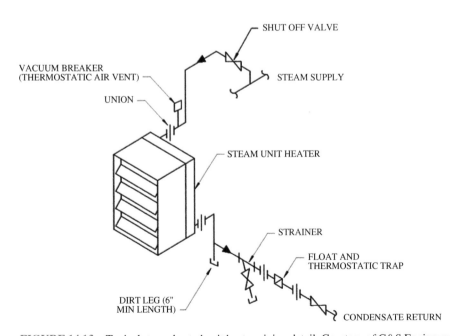

FIGURE 14.13   Typical steam heated unit heater piping detail. Courtesy of C&S Engineers.

15

SPECIALIZED STEAM EQUIPMENT

There are many types of specialized process steam equipment, but we limit our discussion in this chapter to some major types:

- Back pressure turbines
- Steam injection hydro heaters
- Steam super heaters
- Steam discharge mufflers
- Jacketed kettles
- Sterilizers/steam in place (SIP) systems
- Steam reboilers
- Steam to hot water generators

BACK PRESSURE TURBINE

Back pressure turbines (BPT) use high-pressure steam to generate electricity or provide mechanical work (i.e., shaft work). Facilities that can generate high-pressure (>200 psig) steam but use lower pressure steam can employ a back pressure turbine. Those types of facilities often use pressure reducing valves to step down the high-pressure steam to a lower pressure. A back pressure turbine can be used in place of a PRV. Figure 15.1 shows a typical back pressure turbine application.

The use of a BPT is different from a condensing turbine. In a condensing steam turbine (what is normally used to generate electricity in a steam power plant), the steam condenses, is collected and fed back to the steam generator. A BPT does not condense the steam; it just extracts enthalpy by reducing its pressure. The higher the

Process Steam Systems: A Practical Guide for Operators, Maintainers, Designers, and Educators, Second Edition. Carey Merritt.
© 2023 John Wiley & Sons, Inc. Published 2023 by John Wiley & Sons, Inc.

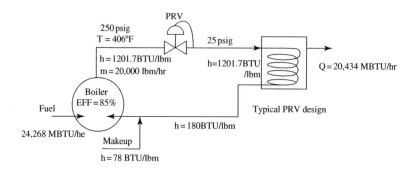

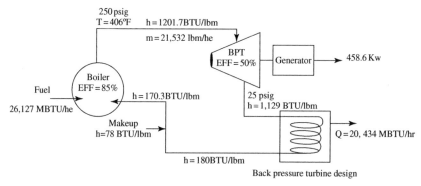

FIGURE 15.1   Back pressure turbine versus PRV. Data from *Combined Heating, Cooling and Power Handbook* [46].

pressure reduction across the BPT, the more energy that can be extracted and eventually converted to electricity or mechanical work. Figure 15.2 shows some general energy conversion capability starting with various inlet pressures.

One can also calculate the amount of added steam flow required to operate the BPT. First, obtain the enthalpy loss data across the BPT and determine the added steam flow at the turbine inlet to account for the loss.

Example 15.1

*A 240 bhp boiler @ 280 psig with 180 F feed water will provide 7683 lb/h steam flow having and enthalpy of 1203 btu/lb. This corresponds to 9,242,649 btu/h total energy output. The enthalpy (from manufacture's data with these inlet conditions) of a BPT exhaust steam pressure of 80 psig will be 1154 btu/lb and have 3.5% moisture @ 416 F. Therefore, the losses across the BPT are as follows: for the moisture fraction, 7683 × 0.035 = 269 lb/h loss (moisture) × 1203 – 416 btu/lb = **211,703 btu/h** and for the steam fraction enthalpy loss, 7683 – 269 × 1203 – 1154 btu/lb = **363,286 btu/h**. So total losses are 211,703 + 363,286 = **574,989 btu/h**. 574,989 btu/h equates to 574,989 btu/h/1203 btu/lb = **478 lb/h** steam loss across the BPT.*

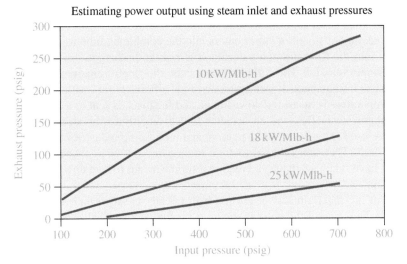

FIGURE 15.2 Estimating power output from a back pressure turbine. Courtesy of the Advanced Manufacturing Office, US Department of Energy [47].

*There is one recalculation we must do to show a more accurate steam loss calcula-tion. The 3.5% moisture will be separated and sent back to the condensate receiver. The energy from the 416 F water will need to be added to the 180 F feed water. Therefore, the 269 lb/h of 416 F water from the BPT will return 416 – 180 × 269 lb/h or **63,484 btu/h**. The condensate energy is raised by this amount and will increase the boiler output by 63,484/416 – 180 + 810 btu/lb = **61 lb/h**. So, the actual boiler output is 7683 + 61 = 7744 lb/h. Consequently, all the subsequent energy balances should be reperformed. This cycle of recalculation can be done several times; how-ever, one recalculation is generally accurate enough.*

From Figure 15.2, this same application of 280 psig steam stepped down to 80 psig through a BPT should yield about 12 kW/Mlb/h. For this example a BPT could yield 12 kW/Mlb/h × 7.744 lb/h = **92.9 kW**. The savings in electrical costs may far out-weigh the cost of generating the extra 478 lb/h steam. To generate 487 lb/h steam with a boiler that is 80% efficient, we need [487 lb/h × (416 F – 180 F) + 810 btu/lb latent heat]/80% efficient = 636,752 btu/h. The cost of electricity generation at $5/ MMbtu is $3.18/92 kW = **$0.034/kW**.

STEAM HYDRO HEATER

A second specialized piece of steam equipment is a hydro heater. A hydro heater is actually a branded name of a steam injector manufactured by the Hydro-Thermal Corporation. Although there are other similar functioning pieces of equipment on the market today, the name hydro heater seems to best fit how to describe the function of this machine. Hydro heaters are internally modulating steam injectors that inject steam with a 90 degree fluid flow design favorable for slurries that require a high shear heating. A liquid or slurry enters into the combining tube where steam is introduced into the fluid through a nozzle. The nozzle discharges steam at very high, often sonic velocity. The turbulent nature of this high velocity discharge enables steam to instantaneously penetrate, disperse, efficiently mix with the slurry or liquid. Temperature is measured downstream and feedback is sent to a controller which modulates steam flow via an actuator linked directly to the steam nozzle plug. They are used extensively in the pulping and biorefining industries to heat up the process fluid. They can be used in food processing and other manufacturing facilities that have steam available and require a simple means to heat up a high flow process stream.

The flow of process fluid and steam in a hydro heater is shown in Figure 15.3. Essentially the flow rate of the process stream is manually controlled by moving an adjustment cone more open or more closed. The hydro heater automatically adjusts the steam flow to maintain temperature of the process fluid; therefore, no external steam control valve is needed. A typical system integration piping diagram is shown in Figure 15.4. The violent mixing helps ensure the process fluid is heated uniformly and can aid in breaking up particulate mixtures. In corn-based ethanol plants, hydro heaters are used to add thermal energy to corn mash slurries and aid in gaining access to the starches inside the corn fiber shell.

There are design considerations that should be used when applying a hydro heater. The steam supply will need a minimum pressure and should be of high quality. Strainers and drip legs should always be utilized in the steam line upstream of the heater. There are process flow and pressure relationships to steam flow and pressure that are required through the hydro heater. Sizing is critical and consultation with the manufacture should occur prior to system design. Unmatched process flow and steam flow can create excessive vibration and noise. When tuned properly, a hydro heater is quiet and operates very efficiently. Sometimes a valve is placed in the discharge of the heater to ensure a slight back pressure is placed on the unit to help balance flows.

STEAM SUPERHEATERS

Occasionally a process heating application may require heating a product to a temperature that makes using saturated steam impractical due to the required very high steam pressure. For those applications an independent hot oil system can be used or a steam superheater may be employed. Hot oil systems use hot oil circulated under low pressure in lieu of steam. They can be expensive. If a steam supply exists in a facility, an alternate solution might be to add a steam superheater to add sensible energy to the existing steam supply. Superheaters can be gas fired, electric, or use waste heat to add energy to the steam supply. Superheated steam will be at the pressure of saturated steam but have a higher temperature. An electric-type superheater is shown in Figure 15.5.

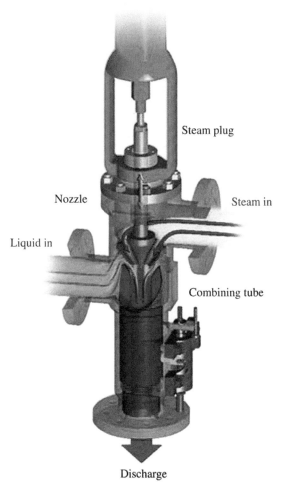

FIGURE 15.3   Hydro heater. Hydro-Thermal Corp.

Example 15.2

The facility where you work has an abundant supply of 125 psig steam; however, a new process is being evaluated that requires a product to be heated to 500 F. What are your heating options? You have determined that the new process will require 750,000 btu/h. You have three options. 1. Purchase a steam boiler rated for >700 psig operating pressure. 2. Purchase a hot oil system rated for >500 F oil service, or 3. Use a steam superheater to raise the enthalpy of the 125 psig steam high enough to provide >500 F temperature steam.

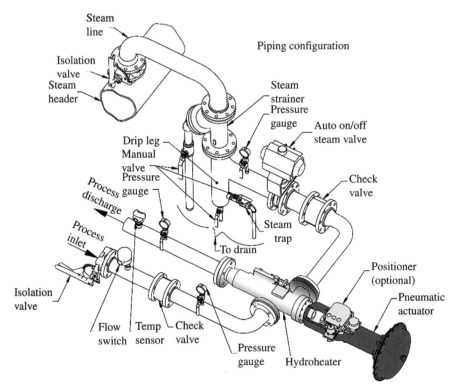

FIGURE 15.4   Piping diagram of installed hydro heater. Hydro-Thermal Corp.

FIGURE 15.5   Electric steam superheater skid mounted with electric boiler. The Fulton Co.

Answer: You wish to investigate how much electrical power usage it will take to superheat the existing steam supply. We know the amount of saturated steam required to yield 750,000 btu/h will be 750,000 btu/h ÷ 871.5 btu/lb (latent heat at 120 psig) = 860 lb of steam. A good place to start is to look at the superheated steam tables in Appendix A and find out the difference in enthalpy of 125 psig saturated steam and superheated 125 psig steam at 550 F. The tables show 125 psig (140 psia) steam to have an enthalpy of 1193 btu/lb and 550 Fand 125 psig superheated steam to have an enthalpy of 1301.3 btu/lb. Therefore, the amount of energy required to superheat the stream is 197 btu/lb. We also can determine the amount of superheated steam by now dividing the 750,000 btu/h by 871.5 btu/lb + 197 btu/lb = 702 lb. The electrical energy required to raise 702 lb/h steam by 197 btu/lb = 138,294 btu/h. We know that 3412 btu = 1 kW; therefore, the amount of electrical energy required per hour for this application is 138,294 btu/h/3412 btu/kW = **40 kW***.*

The evaluation of which option to use to provide 500 F heat transfer fluid temperature would determine the capital and operating costs to add the new superheater versus adding a hot oil system versus the cost of a 700 psig boiler. Both the hot oil and the high-pressure boiler options will have high initial capital costs.

Steam superheating may also occur in a process boiler itself [24]. Larger water tube-type boilers can be designed to pass the steam through stages of exhaust gas heat exchangers to super heat the steam. In some applications feed water may be injected into the superheated steam to control the steam temperature. The added water does not fully de-superheat but flashes to steam and absorbs enough enthalpy to lower the steam temperature to a set point. Steam temperature produced from the boiler can be controlled to fairly narrow temperature limits throughout a variety of boiler firing rates.

STEAM DUMP MUFFLERS

For large process steam systems, it may become necessary to install a steam muffler on the steam dump line. If during a process, the need for steam is suddenly stopped, the excess steam produced will be vented in mass from the steam header to the boiler building exterior. This situation can result of loud high-pitched irritating noise. This is especially true for process boilers located in municipalities where community awareness is high. A steam muffler can be employed to reduce the noise level to acceptable levels. One such application is shown in Figure 15.6. The muffler acts similar to the one found on an automobile exhaust.

The muffler system design must take into consideration the forces created by discharging a high mass of high-pressure steam and must account for the condensate that will likely be created in the muffler itself. Construction should be stainless steel, and the mounting should be approved by a qualified structural engineer.

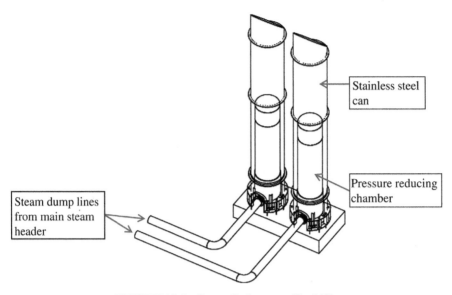

FIGURE 15.6   Steam discharge muffler [48].

JACKETED KETTLES

Jacked kettles are used in food, beverage, and chemical manufacturing. This application uses steam in an interstitial area around a tank or vat to heat the material in the vessel. A typical piping arrangement for a jacketed kettle design is shown in Figure 4.2. Essentially, steam is added to the jacket proportional to the temperature of the material in the vessel. Steam flow and condensate removal rate are critical controls. Product heat up rate and temperature are parameters controlled by the steam supply. Poor steam control can allow uneven heating rates, and poor condensate removal can allow for water logging and delayed heating of the product in the vessel. Means to allow air venting, proper steam control valve, and steam trap sizing are critical.

STERILIZERS

Sterilizers and autoclaves are closed vessels that use steam to kill any biological organisms present on the material inside the vessel. Health care facilities, kitchens, some high-tech manufacturing use sterilizers to provide sterile equipment. Sterilizers generally require high-quality steam or dissolved solids carried over in entrained boiler water will be left on the equipment being sterilized. Not good for any health care or high-tech manufacturing use. A typical sterilizer application is shown in Figure 15.7. This type of steam use almost always results in the condensate being contaminated and discarded. One operational concern is the potential for a very high initial steam demand. When an autoclave is cold the equipment inside will condense

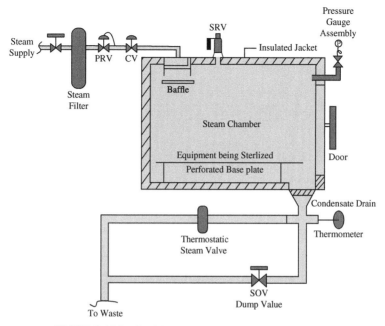

FIGURE 15.7   Sterilizer application Modified from [49].

the steam rapidly and can create enough vacuum to pull in large amounts of steam. An analogy would be the in-rush current to an electric motor when the motor is first turned on. Restricting steam orifices in the supply line can help mitigate the effects of this high initial steam demand.

Sterilization in place (SIP). SIP systems are used extensively in the food, dairy, and biotech industries. SIP is used in conjunction with clean in-place (CIP) systems to provide in situ cleaning and sterilization. Essentially, the CIP systems wash away the residual product material and the steam is injected into lines and vessels to heat up the metal and kill any biologically active organisms. SIP usually follows CIP operations. The steam is injected into tanks via spray balls and held there long enough to kill the microscopic organisms. Temperature reaching 250 F are achievable. The keys to a good SIP are to ensure the system gets enough steam to do the job. Special sanitary stream traps are used and strategically placed temperature sensors are used in a good system design. Similar to steam used in the autoclave application, the resultant condensate is discarded.

REBOILERS

A reboiler is really a steam heat exchanger used to continuously heat a recirculated material through a secondary process like distillation. A typical reboiler setup is shown in Figure 15.8. Product is pumped through a reboiler to a column to allow

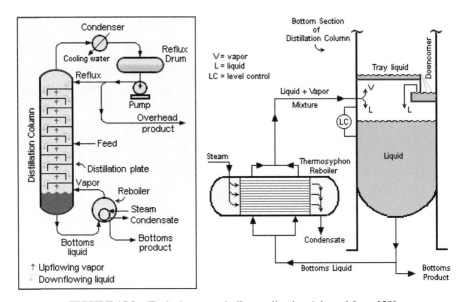

FIGURE 15.8   Typical steam reboiler application Adapted from [50].

evaporation or some other separation process to occur. The reboiler raises the produce temperature enough to allow some form of evaporation to occur. Distillation
of beer and chemical intermediates use a reboiler to add the heat lost during the
evaporation process. Heated product is pumped into the column, pressure is lowered, evaporation occurs, and the cooler solution falls to bottom and pumped back
through the reboiler. Since this is a continuous process, the steam flow required
modulation and almost always done via pneumatic control valves. Reboiler issues
can be poor steam flow control, poor condensate removal, or poor product
circulation. Fouling of reboiler tubes will oftentimes require a system outage to
regain proper heat transfer.

16

STEAM SYSTEM EFFICIENCY/ SUSTAINABILITY

Efficiency is a concept that interests most everyone today mainly due to the fact the boiler room is generally one of the largest energy users in a facility. With the price of energy projected to increase, boilers that use less fuel and produce lower emissions are popular. Increasing the overall efficiency of the steam system only one percent is substantial. Unfortunately, most energy managers fail to look at the entire steam system, rather they focus only on the boiler or just the steam traps. While the boiler efficiency is very important, optimizing an entire steam system should also be considered. This chapter will focus on efficiency of the boiler and the steam system as a whole, with emphasis on understanding what system enhancements can be done to improve efficiency and fuel savings. Likewise, many organizations have an interest in sustainability. Sustainability managers would be well served to fully understand the principles in this chapter. The steam system is an energy hog and every plant should strive to cut energy use and limit emissions.

THE SYSTEM HEAT BALANCE

Before we can discuss efficiency in a steam system, we must first understand energy input and output in all parts of the system. This is called the heat balance and is shown below in Figure 16.1. Fuel energy in the form of chemical or electrical energy converts water to steam. This energy in the steam is transmitted through piping and eventually transferred to the product being heated. In this process, we know from Chapter 3 that radiant, convection, and conduction heat transfer takes place. The key to fuel savings is to maximize heat transfer from the fuel energy to the water and to minimize the heat losses once the steam is formed. Energy managers attempting to employ energy conservation programs should always look at the entire system heat

Process Steam Systems: A Practical Guide for Operators, Maintainers, Designers, and Educators, Second Edition. Carey Merritt.
© 2023 John Wiley & Sons, Inc. Published 2023 by John Wiley & Sons, Inc.

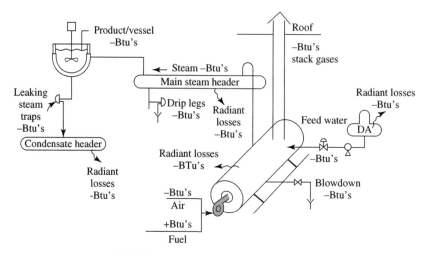

FIGURE 16.1   Steam system heat balance.

balance and first understand the flow of energy into and out of the system before attempting to tackle one piece of equipment.

Boiler Heat Balance

The boiler is where the fuel is converted to steam. Fuel energy is inputted to the boiler and the energy is transferred to the water to make steam. Not all of the fuel energy is captured to make steam. Consequently, some of the energy is swept out of the boiler via the flue gas and blowdown water or through radiant heat losses from the boiler itself. Good fuel combustion and heat transfer of the hot flue gases through the pressure vessel wall to the boiler water will maximize steam production. Furthermore, minimizing blowdown heat losses by incorporating heat recovery systems will help maximize steam production. Radiant heat losses are relatively constant but can be minimized by proper boiler selection and insulation. In addition to the heat losses, some energy is required to heat the combustion air and feed water entering the boiler. Maximizing the air and water temperatures coming into the boiler will also help maximize steam production. A boilers heat balance can be described by

$$\text{Fuel energy in} = \text{steam energy out} + \text{stack energy out} + \text{blowdown energy out} + \text{radiant losses}$$

BOILER EFFICIENCY

Essentially, the boiler efficiency can be stated as the **combustion efficiency**, the **thermal efficiency,** or the **fuel to steam efficiency** [47, 52]. Each is described below and represents a different value. Savvy manufactures will market their boiler with a stated efficiency. One should always ask what type of efficiency the data represents and what were the operating conditions from which the data was derived. The most popular efficiency types used today are combustion and fuel to steam efficiency ratings Here are some of the conceptual principles for these two types of efficiency in a steam boiler:

- Combustion efficiency (theoretical) ≈ burner performance, PV cleanliness, convection heat transfer of flue gas energy to the water side, combustion air temperature, and feed water temperature.
- Combustion efficiency (measurable) ≈ stack temperature and flue gas CO_2, CO, and O_2.
- Boiler fuel to steam efficiency = steam enthalpy (btu/l) × mass flow out (lb/h) ÷ fuel energy input (btu/h).
- Fuel to steam efficiency = combustion efficiency – radiant losses – blowdown losses.

Combustion efficiency: Combustion is the process of mixing and igniting air and fuel to produce a hot usable gas. For natural gas and propane, the chemical formulas are shown below:

$$CH_4 + 2O_2 \rightarrow CO_2 + 2H_2O \text{ Natural gas}$$

$$C_3H_8 + 5O_2 \rightarrow 3CO_2 + 4H_2O \text{ Propane}$$

All fuels have some ideal ratio of fuel to air for perfect combustion. As a rule of thumb, 1 cubic ft of natural gas will require about 10 cubic ft of air for perfect combustion (see Table 13.3). Unfortunately, it is very difficult to obtain perfect combustion and some additional air must be added to ensure all the fuel is consumed. This additional air is called "excess air." Typical excess air levels are in the 15–20% range. Combustion efficiency is an indication of the boiler's ability to burn fuel efficiently and transfer that fuel energy to the boiler water. The amount of unburned fuel and excess air in the exhaust are used to assess a burner's combustion efficiency. Burners resulting in low levels of unburned fuel while operating at low excess air are considered to be very efficient. Scientists and engineers have developed the relationships between excess air, exhaust temperature, and efficiency for a variety of liquid and gaseous fuels. Setting combustion in a boiler to achieve optimum efficiency is therefore of primary interest. Stack gas analyzers can measure oxygen, temperature, and other parameters to ensure the air to fuel ratio is optimized. The relationship between fuel, air, and CO in boiler flue gas is shown below in Figure 16.2. **A rule of thumb is for every 10% excess air beyond what is required for complete combustion will result in a 1% combustion efficiency loss**.

It would be a safe to bet that many boilers operating today are not set up for optimum combustion for all times of the year or over the entire burner firing range. Optimal combustion must be set at all firing rates and should be reset as a minimum for summer and winter seasons. As a burner modulates down, the air and fuel input to the burner are controlled to reduce the burner output. Many older burners in use

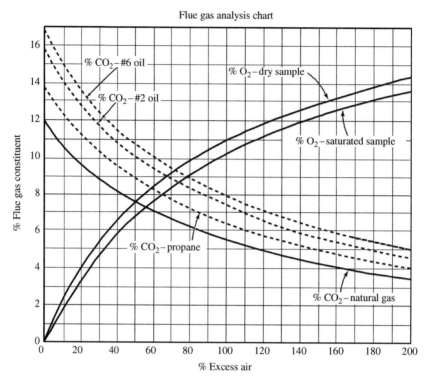

FIGURE 16.2   Flue gas analysis versus excess air. Eclipse Inc.

today employ mechanical linkage systems to connect the air input gate and fuel valve to modulate the burner input. Unfortunately, flow through air gates and fuel valves is not linear with valve position, so it becomes difficult to maintain the fuel to air ratio for optimal combustion at very low burner output (turndown). Consequently, when the boiler load dictates a low burner output, too much or too little excess air is added and combustion efficiency suffers. The development of electric servo controls and PLC-based combustion control to modulate the air and fuel independently have allowed burner manufactures to make burners that have much improved combustion at even low turndown levels. In addition, boiler mounted exhaust oxygen monitors (i.e., O_2 trim) or inlet air density compensation can now provide a constant feedback signal to these PLC combustion controls to further optimize combustion efficiency. Retrofitting an existing boiler with these modern combustion controls is a significant step in improving the system overall efficiency.

Once the fuel is combusted, the energy must be transferred to the boiler water. As we have seen earlier in this book, this heat transfer occurs according to the formula $Q = m \times Cp \times \Delta T$. This expression of heat transfer is directly related to m or surface area, Cp or thermal conductivity of the flue gases, pressure vessel wall and boiler water, and the difference in temperature between the flue gases and the boiler water. Therefore, combustion efficiency is affected by

1. Boilers pressure vessel surface area: the more the better; this is a design feature of the boiler. Turbulators, Alufer technology (see Figure 16.3), and econo-mizers essentially increase the heat transfer surface area.
2. Thermal conductivity of
   a. Flue gases: need turbulent flow; low fire conditions can result in laminar flow through the boiler furnace which will reduce convective heat transfer.
   b. Pressure vessel wall: cleanliness is paramount; scaling, soot, or sludge buildup will lower thermal conductivity.
   c. Water: thermal conductivity of the water is generally not an issue; boiling helps convection heat transfer.
3. Temperature difference (flue gas temperature − boiler water temperature): the higher the ΔT, the more heat transfer. Low-pressure boilers are more efficient than high-pressure boilers simply because the boiler water temperature is lower, which results in a higher ΔT.

The combustion efficiency charts in Appendix A show the stack losses of a fuel-fired boiler using the exhaust CO_2 levels and net stack temperature. These charts have been developed for both gas and oil-fired units.

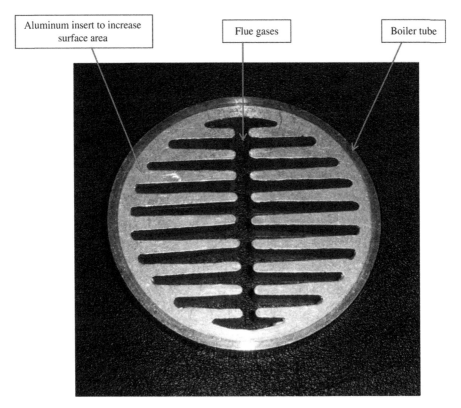

Aluminum insert to increase surface area

Flue gases

Boiler tube

FIGURE 16.3 Cross section of Alufer® heat transfer technology. Developed by Cleaver-Brooks, Inc.

Example 16.1

From the combustion efficiency table in Appendix A, a natural gas-fired boiler operating at 6.0% stack gas CO_2 and having a net stack temperature of 300 F (net is actual – ambient) will have 20.4% stack losses or 79.6% combustion efficiency. If turbulators are added to the boiler tubes and caused stack temperature to decrease to 240 F net, then the boiler would have 18.2% stack loses or have 81.8% combustion efficiency. Likewise, if a burner control modification could be done that would allow the exhaust CO_2 levels to increase to 7% then at 240 F net stack temperatures, the stack losses would be 17.1% and the combustion efficiency rises to 82.9%.

A typical combustion efficiency curve versus firing rate is shown below in Figure 16.4. Notice the efficiency is lower at the low and high fire rate extremes. The reason the loss of efficiency at the low end is due to the poor air to fuel ratio control and laminar flue gas flow in the furnace chamber. At high fire rate, the boiler pressure vessel surface area becomes the limiting factor for efficiency. The optimum boiler efficiency generally occurs in the 20–80% output range.

Thermal efficiency is defined as the amount of energy from the combustion of fuel that is transferred to the boiler water. It does not account for radiant losses of the boiler. It is a measure of the ability of the boiler pressure vessel to absorb the heat energy from the hot flue gases. For the layperson, the thermal efficiency concept is not very useful.

Fuel to steam efficiency is the true measure of a boilers' efficiency. It will correlate the amount of input fuel energy to the steam energy going out. It accounts for combustion efficiency, heat transfer to the boiler water, all radiant heat loss and heat loss due to blowdown discharges. Measuring this precisely can only in a laboratory type setup with all operating conditions held constant. We can, however, get a fairly

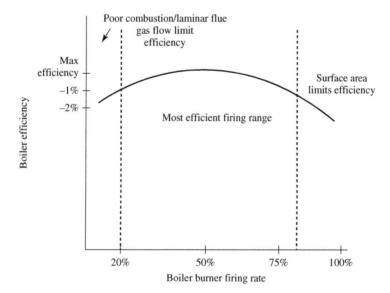

FIGURE 16.4   Typical combustion efficiency versus boiler firing rate.

accurate determination of this efficiency rating by measuring the amount of fuel in, water in, or steam flow out under steady-state conditions. Obviously the feed water temperature and steam pressure needs to be held constant during this determination. The radiant heat loss from a boiler can be obtained from the boiler manufacture or calculated using boiler skin temperature and determining total surface area. Radiant heat loss is a function of the boiler pressure, not output. Therefore, a boiler at a given pressure will have the same radiant heat loss whether it is at low fire or high fire. The example below demonstrates that radiant losses influences boiler efficiency greater when operated at low fire than at high fire. This consequence is significant when you consider that engineers tend to size a boiler for worst-case demand and consequently the boiler may operate at low fire a significant amount of time. It is much better to design two smaller boilers in lieu of one large unit for this reason. By shutting down one of the two boilers in low demand periods make good sense. Another overlooked heat loss from a boiler is the loss of heat each time a boiler is started and shut down. Pre- and postpurge air flow will carry internal boiler heat energy right up the stack. A steam system with poor turn down will cycle excessively through these purges at low steam demand periods.

Example 16.2

A boiler manufacture indicates that their 200 hp horizontal firetube steam boiler will have an rated output of 5,520,000 btu/h and a radiant heat loss of 18,500 btu/h when operated at 100 psig. What is the boiler efficiency at low, medium, and high fire given the typical combustion data below and considering the burner has a 10:1 turndown:

Low fire: Stack temperature is 280 F net and CO_2 is 6.0 ppm.
Mid fire: Stack temperature is 310 F net and CO_2 is 6.0 ppm.
High fire: Stack temperature is 350 F net and the CO_2 is 6.0 ppm.

*Answer: Boiler efficiency can be represented as the combustion efficiency minus radiant heat losses. From the combustion efficiency tables in Appendix A, at high fire, the efficiency loss due to radiant heat loss is 18,500 btu/h/5,520,000 btu/h = 0.34%. Furthermore the combustion efficiency is shown to be 77.8%. Therefore the overall efficiency is 77.8 – 0.34% = **77.5%**. Similarly at mid fire the combustion efficiency is 79.3%. The efficiency loss due to radiant heat loss is 18,500/50% × 5,520,0,000 btu/h or 0.67%. Therefore the overall efficiency at mid fire is 79.3% – 0.67% = **78.6%**. At low fire the boiler will have only 1/10th the output or 5,5200,000/12 = 520,000 btu/h. Combustion efficiency is 80.4% and the efficiency loss due to radiant losses is 18,500 btu/h/520.000 btu/h = 3.6%. At low fire the overall efficiency is 78.6% – 3.6% = **76.8%**.*

STEAM SYSTEM EFFICIENCY

There are several design enhancements that can be utilized to help improve efficiency. Many are outlined in Table 16.1.

TABLE 16.1   **Steam System Performance Improvements**

Opportunity	Description
Generation	
Minimize excess air	Reduces the amount of heat lost up the stack, allowing more of the fuel energy to be transferred to the steam
Clean boiler heat transfer surfaces	Promotes effective heat transfer from the combustion gases to the steam
Install heat recovery equipment (feedwater economizers and/or combustion air preheaters)	Recovers available heat from exhaust gases and transfers it back into the system by preheating feedwater or combustion air
Improve water treatment to minimize boiler blowdown	Reduces the amount of total dissolved solids in the boiler water, which allows less blowdown and therefore less energy loss
Recover energy from boiler blowdown	Transfers the available energy in a blowdown stream back into the system, thereby reducing energy loss
Add/restore boiler refractory	Reduces heat loss from the boiler and restores boiler efficiency
Optimize deaerator vent rate	Minimizes avoidable loss of steam
Distribution	
Repair steam leaks	Minimizes avoidable loss of steam
Minimize vented steam	Minimizes avoidable loss of steam
Ensure that steam system piping, values, fittings, and vessels are well insulated	Reduces energy loss from piping and equipment surfaces
Implement an effective steam-trap maintenance program	Reduces passage of live steam into condensate system and promotes efficient operation of end-use heat transfer equipment
Isolate steam from unused lines	Minimizes avoidable loss of steam and reduces energy loss from piping and equipment surfaces
Utilize back pressure turbines instead of PVRs	Provides a more efficient method of reducing steam pressure for low-pressure services
Recovery	
Optimize condensate recovery	Recovers the thermal energy in the condensate and reduces the amount of makeup water added to the system, saving energy and chemicals treatment
Use high-pressure condensate to make low-pressure steam	Exploits the available energy in the returning condensate

[25] / U.S Department of Energy / Public Domain.

One simple enhancement is incorporating a **standing pilot**. In a normal boiler burner control scheme, the pilot flame will extinguish after the main flame lights off. This enhancement allows the pilot to remain lit even after the operating pressure set point is reached. With the pilot flame lit, and the main flame idle, a call for heat signal received from the pressure controls will allow the burner to go directly to main flame without a prepurge cycle. The standing pilot option is only allowed for boilers <2 million btu input (CSD-1 code). For small boilers employing on/off burners, this can be a good enhancement. An extra pressure switch is added to provide a signal to shut off the pilot flame if the steam pressure starts to creep toward the high-pressure alarm set point.

Combustion air preheating. Preheating the combustion air can gain some increase in boiler efficiency. When air is mixed with the fuel in the burner and ignited, it will produce heat energy. If less energy from combustion is needed to heat the incoming air, then more energy is to be transferred to the boiler water. To calculate how much energy savings, we need to determine the amount of fresh make up air and the new preheated air temperature.

Example 16.3

*A 200 hp natural gas-fired boiler operated at rated output will require about 8625 ft³ input of natural gas input per hour. With 15% excess air we would need 129,375 ft³/h fresh air for good combustion. From a HVAC Handbook [51] the heat capacity of moist air is about 0.24 btu/lbm-F and the density of this air at 70 F is about 0.07 lbm/ft³, then the energy to raise 129,375 ft³ of combustion air per hour would be about 2173 btu/F. Therefore, if we raise the inlet combustion air temperature 100 F, we could save about 2173 btu/F × 100 F = 217,300 btu/h. This is about a 2.5% increase in efficiency. **A good rule of thumb for combustion air preheating is a 40 F increase in combustion air temperature will yield about a 1% increase in boiler combustion efficiency**.*

Economizers. An economizer is nothing more than an air to water or air to air heat exchanger mounted in the exhaust stack of a steam boiler. The hot flue gases flow through the economizer and energy is transferred to a supply of water. A typical flow diagram is shown below in Figure 16.5. Where economizers are installed to preheat feed water, highly deaerated feed water supply is essential. Because oxygen pitting is the most common cause of economizer tube failure, the supply feed water should be treated with an oxygen scavenger type chemical. Installation of flue gas economizers make sense for all larger boilers (i.e., >150 hp), but may not be economical for smaller boilers. The boiler feed water can be the water supply, but care should be taken not to lower the flue gas temperature to less than 200 F or flue gas condensing will take place. Condensed flue gas is corrosive and will destroy the boilers exhaust system. **A general rule of thumb is that for every 40 F you lower the boiler exhaust flue temperature, you can expect about 1% efficiency gain**.

Example 16.4

A 67,000 lb/h natural gas-fired water tube boiler operating at 125 psig utilizes a flue gas economizer to preheat feet water as shown above. The cost to install this economizer is $250,000. What is the payback period if the cost of natural gas is $6 per million btu?

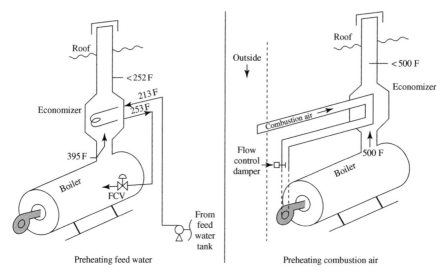

FIGURE 16.5   Boiler stack economizer system integration.

Answer: The flow diagram above shows that the flue gas temperature is lower from 395 to 252 F, and the feed water temperature is raised from 213 to 253 F. We know that the boiler is operating at about 67,000 lb/h which is essentially the feed water flow. The energy extracted by the economizer is calculated as 67,000 lb/h water flow × 1 btu/lb × 253 – 213 F = 2,680,000 btu/h. And the cost of natural gas is $6/ MMbtu, so the hourly savings is 2.68 × 6 = $16.08/h. Extrapolated to an annual savings means $16.08/h × 8760 h/yr = $140,861/yr. The simple payback would then be $250,000/$140,861 = 1.8 years.

Economizers can also be used to preheat combustion air. Although less energy is extracted using this application compared to preheating feed water, it will increase the boilers efficiency. These types of economizers are normally found in applications where feed water preheating is not required or desired. Economiozer and other heat recovery technologies can be integrated into a steam system as shown in Figure 16.6.

Blowdown heat recovery units. In addition to proper blowdown practices, including the use of automatic blowdown control, reducing cost and heat loss associated with boiler blowdown can also be achieved through recovering the heat/energy in the blowdown. The blowdown water as it leaves the boiler will have the same temperature and pressure as the boiler water. Before this high-energy waste is discharged, the heat in the blowdown water can be recovered with a flash tank, a heat exchanger, or the combination of the two. Any boiler with continuous surface water blowdown exceeding 5% of the steam generation rate is a good candidate for blowdown water heat recovery. Blowdown heat recovery economics are shown in Figure 16.7

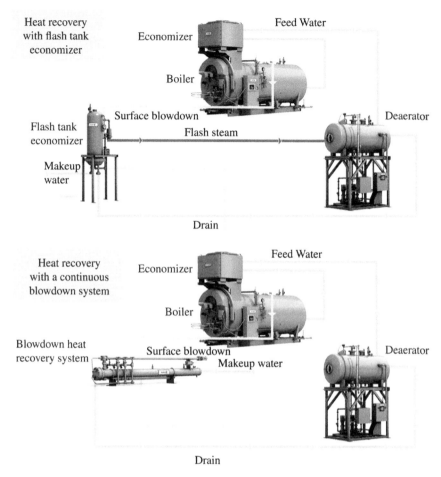

FIGURE 16.6   Integrated heat recovery systems. Courtesy of Cleaver-Brooks, Inc.

Example 16.5

You are installing a 300 hp, 120 psig horizontal firetube boiler that will have 75% returns, a makeup water conductivity of 250 uS/cm². Assuming this boiler will run at 80 % capacity, how much energy can be recovered from blowdown discharges if we use ABMA guidelines and keep the boiler water conductivity at 3000 uS or less.

Answer: The boilers mass flow of blowdown water can be calculated using the formula BD (lb/h) = effective evaporation rate (ER)/1 – cycles of concentration (COC), and the effective evaporation rate is the rated output × the capacity × the amount condensate lost. Since the condensate lost equals 1 – percentage condensate return, our application has 1 – 0.75 or 25% condensate lost. The effective evaporation rate is then 300 hp × 34.5 lb/h/ BHP × 80% capacity factor × 25% condensate lost. We can see the ER = 2070 lb/h. Next we need to calculate the cycles of concentration or COC, which is the ratio of boiler

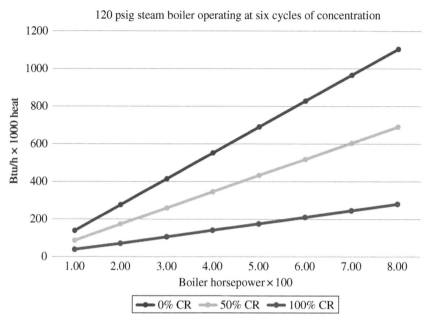

FIGURE 16.7   Blowdown heat recovery economics.

water to the makeup water conductivity. For our application the COC equals 3000 uS/250 uS = 12. We now can calculate the blowdown rate to be 2070 lb/h/1 – 12 = 188 lb/h. From the steam tables we can see that 120 psig boiler water has an enthalpy of 321 btu/ lb. If we use the blowdown water to make useful 125 F hot water, then the enthalpy of 125 F water is only 93 btu/lb. We then can see the amount of recoverable sensible heat for this application is 321 – 93 (btu/lb) × 188 lb/h = 42,864 btu/h. The 42 K btu/h represents an ideal heat transfer and in reality the value is likely closer to 80% of this value or 34,300 btu/h. Even at 34 K btu/h, this is almost 300 million btus recovered per year. The decision to install a blowdown heat recovery system would depend on the cost and how long it would take to recover enough btus to pay for the installation.

Boiler Internal Cleanliness

The boiler internal cleanliness affects heat transfer from the boiler furnace to the boiler water through the pressure vessel wall. Scaling and fouling lower cleanliness can significantly reduce boiler efficiency. Table 3.2 shows the thermal efficiency loss due to scaling and sooting. Weekly boiler stack temperature monitoring and semiannual boiler internal inspections are the best way to monitor boiler internal cleanliness. Sooting is a consequence when improper (to rich) combustion occurs and the furnace side of the pressure vessel gets coated with a layer of carbon. Sometimes the only way to remove the soot layer is by manual scraping or brushing to return the surface to as new cleanliness. Scale can be soft or hard scale. Scale is best removed via chemical cleaning by the professionals who have experience with chemical cleaning. Figure 16.8 shows typical chemical cleaning results.

Steam Delivery System Efficiency

The steam delivery system efficiency can impact the system overall efficiency. Radiant and convective heat losses from hot surfaces can add up quickly if those hot surfaces are not insulated. Table 16.2 shows the radiant heat loss from uninsulated steam lines in still air. In addition, the *North American Combustion Handbook* lists radiant and convection heat losses from a 200 F surface to range from 225 to 565 btu/ft^2-h [52]. The lower value is for horizontal surfaces with no air flow and the high value is for vertical cylinders with high air flow. **In general, a rule of thumb uses a value of about 3 btu/ft^2-ΔF/h to determine heat losses from uninsulated pipes and other hot surfaces**.

Example 16.6

If we use this rule of thumb, a 4 inch steam line uninsulated carrying 125 psig steam located in a room that is 90 F will lose about 1.325 ft^2/ft × 3 btu/ft^2-ΔF (353–90)/h = 1045 btus per foot of pipe per hour. The high rule of thumb valve accounts for air movement around the hot pipe which results in higher convective heat losses. Table 16.2 can be

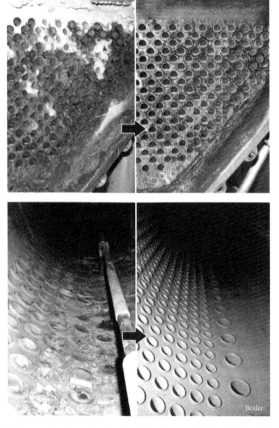

FIGURE 16.8 Before and after chemically cleaning scaled/fouled heat exchange surfaces. Apex Engineering Products Corporation.

used to determine the heat losses from piping with lower surface temperature; however, at higher temperatures the rule of thumb should be used. The heat loss will dictate that all hot surfaces should be insulated.

Steam main drip legs trap and remove any condensate in the header. If the steam traps leak by, steam can be short cycled back to feed water system. Since the drip legs are often not fully insulated, the more mass flow through them will result in higher energy losses. Cleanliness also applies to the steam using equipment. The equipment cleanliness affects the rate of heat transfer of the steam to the product and poor heat transfer to the product may create high boiler cycle times.

Condensate and Feed Water System Efficiency

Losses in this part of the steam system are related to leaking steam traps and heat losses due to uninsulated hot surfaces. Efficiency due to leaking steam traps is well documented. Table 16.3 shows the amount of steam that a leaking trap can inadvertently discharge. Some means of leak detection should be employed and a comprehensive preventative maintenance program adopted that monitors and replace leaking steam traps. Uninsulated feed water tanks, condensate tanks and piping will all allow heat energy to escape the steam system and contribute to system inefficiency.

TABLE 16.2 Heat Loss From Uninsulated Pipe in Still 70 F Ambient Air

Diameter of Pipe (inches)	Temperature of Pipe (°F)					
	100	120	150	180	210	240
	Heat Loss per Lineal Foot of Pipe (btu/h)					
1/2	13	22	40	60	82	106
3/4	15	27	50	74	100	131
1	19	34	61	90	123	160
1-1/4	23	42	75	111	152	198
1-1/2	27	48	85	126	173	224
2	33	59	104	154	212	275
2-1/2	39	70	123	184	252	327
3	46	84	148	221	303	393
3-1/2	52	95	168	250	342	444
4	59	106	187	278	381	496
5	71	129	227	339	464	603
6	84	151	267	398	546	709
8	107	194	341	509	697	906
10	132	238	420	626	857	1114
12	154	279	491	732	1003	1305
14	181	326	575	856	1173	1527
16	203	366	644	960	1314	1711
18	214	385	678	1011	1385	1802
20	236	426	748	1115	1529	1990

Data from *North American Combustion Handbook* [11].

TABLE 16.3 Steam Flow Through Orifices Discharging To Atmosphere

Orifice Diameter (inches)	Inlet Pressure (psig)											
	2	5	10	15	25	50	75	100	125	150	200	300
1/32	0.31	0.47	0.58	0.70	0.94	1.53	2.12	2.70	3.30	3.90	5.10	7.40
1/16	1.25	1.86	2.30	2.80	3.80	6.10	8.50	10.80	13.20	15.60	20.30	29.80
3/32	2.81	4.20	5.30	6.30	8.45	13.80	19.10	24.40	29.70	35.10	45.70	67.00
1/8	4.50	7.50	7.40	11.20	15.00	24.50	34.00	43.40	52.90	62.40	81.30	119.00
5/32	7.80	11.70	14.60	17.60	23.50	38.30	53.10	67.90	82.70	97.40	127.00	186.00
3/16	11.20	16.70	21.00	25.30	33.80	55.10	76.40	97.70	119.00	140.00	183.00	268.00
7/32	15.3	22.9	28.7	34.4	46.0	75.0	104.0	133.0	162.0	191.0	249.0	365.0
1/4	20.0	29.8	37.4	45.0	60.1	98.0	136.0	173.0	212.0	250.0	325.0	477.0
9/32	25.2	37.8	47.7	56.9	76.1	124.0	172.0	220.0	268.0	316.0	412.0	603.0
5/16	31.2	46.6	58.5	70.3	94.0	153.0	212.0	272.0	331.0	390.0	508.0	745.0
11/32	37.7	56.4	70.7	85.1	114.0	185.0	257.0	329.0	400.0	472.0	615.0	901.0
3/8	44.9	67.1	84.2	101.0	135.0	221.0	306.0	391.0	478.0	561.0	732.0	1073.0
13/32	52.7	78.8	98.8	119.0	159.0	259.0	359.0	459.0	559.0	659.0	859.0	1259.0
7/16	61.1	91.4	115.0	138.0	184.0	300.0	416.0	532.0	648.0	764.0	996.0	1460.0
15/32	70.2	105.0	131.0	158.0	211.0	344.0	478.0	611.0	744.0	877.0	1144.0	1676.0
1/2	79.8	119.0	150.0	180.0	241.0	392.0	544.0	695.0	847.0	998.0	1301.0	1907.0

Data from Spirax Sarco [16].

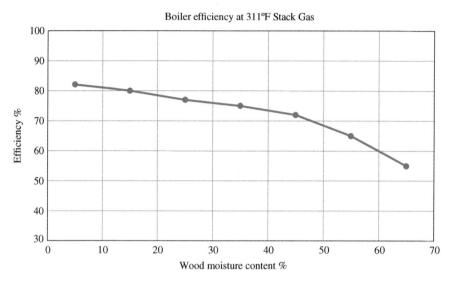

FIGURE 16.9   Biomass water content versus boiler efficiency. Data from *Power Journal*, Vol. 158 [53].

Example 16.7

A feed water tank is 24″D × 48″L made from carbon steel holds boiler feed water at 200 F. How much heat energy is lost from this tank each hour if the boiler room is kept at 90 F?

Answer: From Table 3.4 the heat losses from a 200 F vertical steel surface range from 240 to 565 btu/h. The range accounts for heat loss relative to ambient still air and well-circulated air. The heat loss can be calculated by simply determining the surface area of the tank and multiplying the area by the right heat loss valve. Therefore this tank will have a surface area of $2 \times (\Pi \, r^2) + 4 \times \Pi \, d = 2 \times (3.14 \times 1$ ft) + 4 ft × 3.14 × 2 ft = 31.4 ft². Using a conservative heat loss rate of 300 btu/ft²/h, then the heat loss from this tank is 300 btu/ft²/h × 31.4 ft² = 9420 btu/h. If this system is operated 8 h a day, then the daily heat loss is 75,360 btu/day.

Biomass Fuel Water Content Reduction

This enhancement is more of a system management issue. Biomass boilers, especially those that burn wood chips, can benefit greatly if the fuel they burn is dry. This is an important issue for energy calculations used for determining if a CHP (combined heat and power) using wood chips is feasible. Any moisture contained in the wood chips will be evaporated during combustion. The high latent heat requirement to evaporate water robs combustion energy that could be used to heat water in the boiler. The effect of biomass water content on boiler efficiency is significant and shown in Figure 16.9.

17

SHUTDOWN, STARTUP, INSPECTION, AND MAINTENANCE

So far, this book has focused mainly on steam system design. This chapter is dedicated to explaining good operational, inspection, and maintenance practices that will ensure the boiler and all other system components perform reliably and for many years.

SHUTDOWN AND STARTUP PRACTICES

Smooth startup of a steam system actually is very much related to how the system was shut down. Correctly shutting down a steam system will help ensure a smooth and safe startup. Boilers can be shut off at any firing rate without damaging the pressure vessel; however, residual heat stored in the boiler will cause the boiler to continue to steam after the burner abruptly shuts down. Consequently, if a process steam valve closes rapidly and forces a system shutdown when the boiler is at high fire the safety valves might lift. The boiler standard operating procedures (SOP) should have guidance to monitor boiler pressure and allow for manually relieving steam pressure by dumping steam. Large steam systems should employ a dump valve that will relieve the steam header pressure. Dump valves are usually placed in a branch off the steam header close to the boiler room but exhaust outside. The dump valve can be manually or automatically actuated and should relieve at a pressure at least 10 psig below the lowest safety relief valves (SRV) set point. A general design for a dump valve is shown in Figure 17.1

When a steam system is shutdown, work orders should be written for any problems that may have caused the steam system shutdown. This seems like common sense, but the boilers are usually support equipment to a process and maintenance issues related to the boilers can take a backseat to fixing process-related equipment.

Process Steam Systems: A Practical Guide for Operators, Maintainers, Designers, and Educators, Second Edition. Carey Merritt.
© 2023 John Wiley & Sons, Inc. Published 2023 by John Wiley & Sons, Inc.

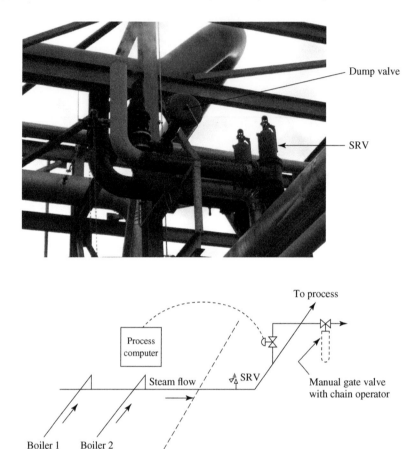

FIGURE 17.1   Steam header dump valve application. Courtesy of Sunmarks, LLC.

Furthermore, anytime you can take a steam system out of service in a controlled manner (versus a boiler trip), you should use the shutdown to check the boiler safeties like the auxiliary low water cutoff function or combustion air switch checks.

Once a boiler shuts down, it will cool and contract. **Never completely isolate a hot boiler pressure vessel until it has cooled**. Isolating a hot boiler can create a significant vacuum in the pressure vessel while the boiler is cooling off. It is good practice to open a vent valve until the boiler is >120 F. One way to help cool off a hot boiler is to manually keep the combustion air fans on with the burner off and let the cool ambient air suck the heat out of the boiler. Even a big boiler can be cooled off completely in a few hours this way. Likewise, the cooling and contraction in the steam lines after a shutdown can cause large volumes of condensate to collect in the steam piping. This condensate must be removed prior to repressurizing the steam header, or water hammer will likely occur.

Generally steam system startup should be slow and methodical. Warm-up means burner light off to steaming up to a few pounds of steam pressure. The reason for a slow startup is to mitigate the effects of thermal stress. When a cold boiler is started up, heat may be absorbed by the pressure vessel unevenly. As metal heats up, it expands and uneven heating causes uneven thermal expansion. This can create significant stress on the pressure vessel metal. Therefore, you should add heat slow enough to allow the heat to migrate to all areas of the boiler vessel evenly. Water circulation inside the pressure vessel, caused by convection between hot and cooler water, will help spread the heat evenly. Since this convection is slow, time is required to allow convection to work. A good practice is to start and stop the boiler on low fire increments of five minutes on and five minutes off until you produce steam.

This prolonged startup period is sometimes called "**soaking**." Obviously, larger boilers will take more time to get operational. A rule I have used is to allow 30–60 min warm up for any boiler system smaller than 100 hp and 30 min per 100 hp after that. A 1000 hp boiler should take at least 5.5 h to warm up properly. The entire steam piping should be warmed up the same time the boiler is soaking. This can be accomplished by keeping the steam stop valve open and allowing the entire header to come up to pressure and temperature evenly. For large steam systems, the header drip legs should be opened and draining during this prolonged startup. Once steam is being produced in the boiler the main header drains can be re isolated. Once the boiler is steaming the firing rate should be manually adjusted slowly (5 psig/2–3 min) until the boiler is within 10 psig of the desired operating pressure. At this point the boiler controls can be placed in automatic mode and allowed to meet system demands.

Actions taken right after a boiler startup should include frequent blowdowns and initiation of chemical feed systems. During the shutdown and startup sequence it is not unusual to have a lot of corrosion products stirred up. Frequent blowdowns will get these contaminants out of the boiler. Likewise, it is good practice to reinitiate chemical feed, especially dispersants if used to help keep the corrosion products suspended.

BOILER SAFETY CHECKS

During boiler operation, the low water alarm/cutout should be checked daily. During this test, an override test switch should be used to avert shutting the boiler down. The test is simple, energize the low water override and drain the water column to verify the **low water alarm cutoff** signal occurs. This test should be recorded on a daily log sheet and kept near the boiler. If the low water signal does not occur on a low water condition, then repeat the test. If it fails again, then initiate a work order to remove the boiler from service and investigate the problem. Repeat tests should be annotated in the daily log. Obviously if the daily test requires frequent retests, then an investigation should commence as soon as possible. The purpose of the daily test is to have confidence that the boiler will actually shut off in a low water condition. Figure 17.2 shows the typical steam boiler safeties systems.

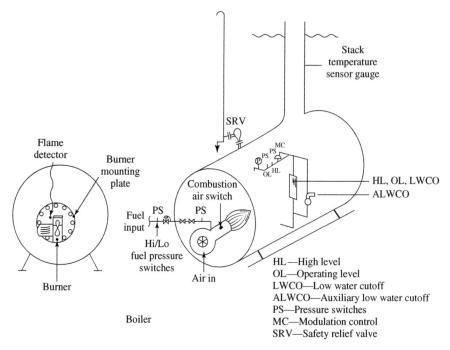

FIGURE 17.2 Boiler safeties.

During nonoperational periods the high and low water alarms can be checked by placing the feed water pump in "hand" and manually overfilling the boiler until the high water alarm activates. The water level can be drained slowly until the first low water cutoff level is achieved. A burner lockout signal should be observed. Refill the boiler and verify the burner lock out will reset. Next redrain the boiler until the second low water cutoff level is achieved. This level should lock out the burner and require a manual reset at the boiler. These checks should also be done whenever the boiler is shutdown for maintenance and performed prior to restart.

The importance of this part of the boiler safety verification cannot be overemphasized. Low water boiler explosions have killed or injured thousands of people over the last century and a half. Fortunately we now have codes that have tremendously reduced this potential hazard. If you find the low water circuitry tampered with or altered from the factory design, you should immediately notify the owner and have the level control system integrity restored prior to completing more work on the boiler.

A second safety feature of a boiler is the **high-pressure cutoff** switches, and these should be periodically checked similar to the low water cutouts. Checking these is a little more difficult. Generally the operating pressure switches need to work for the boiler to operate properly. Essentially the operating switches start and stop the burner when certain pressures are achieved. The highpressure cutout switch should never be checked online unless it is set 90% or lower than the safety relief valve set point. To check to high-pressure switch online, override the operating pressure switch and allow the pressure to rise until the high-pressure cutout activates. This switch requires a manual reset to allow the burner to refire. A safer way to check the high-pressure cutoff switch is to remove the switch from the boiler and bench or field test it to verify the switch will activate at the set pressure. This bench test should be performed at least once per year.

Other safety switches include the **combustion air switch** and **fuel pressure switches**. The combustion air switch senses air pressure in the burner and will not allow the fuel valves to open if there is not enough combustion air to support combustion. This switch is checked every time you fire the burner, so bench testing is not required. Gas pressure switches include high and low gas pressure switches. They are set for each burner/boiler. These switches sense gas pressure upstream and downstream of the main gas valves. They ensure there is high enough gas pressure to support combustion and not so high to prevent the burner from overfiring beyond its design limits. The only way to safely check these switches is to remove them and bench test them. They should be tested at least semiannually. Other switches that should be checked and adjusted periodically include the limit switches on the air inlet dampers. These limit switches allow the burner to cycle through the pre- and postpurge evolutions. Each time the burner lights off, these limits are verified; therefore, only adjustments to ensure reliability need to be performed.

The **SRVs** should also be checked annually. ASME code requires all power boilers (ASME Section I boilers) verify the SRVs will perform their intended function at least once per year. The verification can be done in situ by allowing the boiler pressure to increase to the SRV set point and observe the SRV's lift. This method has one drawback. If the SRV's do not reseat tight after they lift, the boiler may have to be taken out of service for SRV repair. An alternate verification method is to remove the SRVs and bench test the valves. The boiler must be taken out of service to accomplish the task. A good practice is to have a spare set of SRVs that can be installed in place of the testable SRVs. Any time an SRV is removed from a boiler, the boiler connection opening should be inspected and verified there is no obstruction. This inspection and SRV test results should be recorded and kept on file. An SRV inspection and testing form is included in Appendix B. Although SRV testing is mandated at least once per year, it is good practice to increase the testing frequency if a trend of SRV test failures develops.

MAINTENANCE AND INSPECTION PRACTICES

Maintenance on a steam system can be classified into one of three types: corrective, preventative, and predictive maintenance. **Corrective maintenance** is any work done on the system that is required to make the system operational. In general, work orders are submitted to the maintenance department to prioritize and plan. Maintenance workers need to take advantage of every steam system outage and be prepared to mobilize quickly if an outage occurs unexpectedly. It is a good idea to have spare manway and door gaskets, sight glasses, pump seals, gauges, level probes, safety switches, and burner components on hand as spare parts all the time. Feed water pump alignment, control valve timing, combustion analysis, and controls reintegration are critical steps that should be taken after invasive maintenance has been performed on the associated steam system equipment. Steam and water leaks should be addressed as soon as practical. Temperatures around the burner plate where it attaches to the boiler shell should be monitored for temperature at least weekly to ensure refractories are intact. Remember to use lockout tagout procedures before invasion work is performed.

Preventative Maintenance is any work done on the system that will help reduce system downtime. Changing pump seals before they impact system operation, adjusting operating controls, adjusting burners, testing safety valves, cleaning level probes and boiler tubes, checking steam traps and fan bearings, cleaning strainers and chemical feed quills are a few examples of preventative maintenance activities. Most of this kind of maintenance is done at opportunistic times when the steam system is out of service. The frequency of preventative maintenance is generally suggested by the equipment's manufacture; however, corrective maintenance history will help determine the true frequency. All major steam system components should be put into a database where maintenance records can be recorded and trended. Where redundant systems like feed water pumps are utilized, a schedule that rotates the equipment needs to be developed and followed.

The next level of maintenance is **predictive maintenance**. Predictive maintenance is not widely practiced but is gaining recognition among companies that strive for a very strong maintenance program. Predictive maintenance is really a refined version of preventative maintenance. It uses performance data and equipment failure history to predict when a component will fail. Consequently, the preventative maintenance program can be adjusted to prevent or at least minimize unexpected system outages. A predictive maintenance program usually requires an engineer or systems type person to invest a fair amount time gathering data. In addition, most steam systems will need to have additional gauges and sensors installed to allow for data display. An example of predictive maintenance would be recording and trending a pump discharge pressure reading to determine when the pump should be rebuilt. Likewise, recording boiler stack temperature daily is a simple way to track combustion efficiency and how often the boiler tubes should be cleaned. Plotting condensate line temperatures downstream of steam traps or monitoring fan bearing vibrations can provide valuable indication of degraded performance prior to complete failure. Since a steam system is usually a support system to a process, keeping that system online and minimizing downtime is highly desirable. The water treatment professional should be able to trend water chemistry data and predict scaling potential. Corrosion coupons and other side stream monitors can provide strong indication of what is happening inside the steam system.

Inspections

System inspection and predictive maintenance are like ice cream and apply pie. This section will answer what should you inspect and what should you look for during the inspection. Inspections can be daily, monthly, or when the system is shutdown. Inspections are both external and internal. External inspections include observations to identify leaks, equipment material condition, temperatures, water levels, fuel pressure, burner flame color and pattern, chemical feed tank levels, and boiler water chemistry. Internal inspections include looking for scaling, fouling, corrosion, and overall material condition. The internal furnace area of a typical water tube boiler is shown in Figure 17.3

Daily inspections include fuel inlet pressure, flame color and pattern, boiler and DA water clarity, DA temperature and pressure, steam and water leaks in the entire steam system, and boiler water chemistry. The list above seems like a long list, but these items should be checked and recorded at least once per day. An example of a daily log sheet is located in Appendix B. Other important daily inspections include observing water level stability and clarity. Turbid or reddish colored water in the

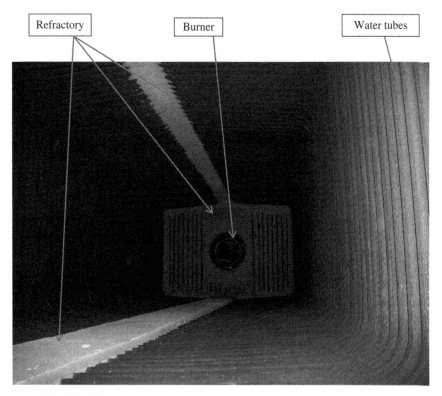

FIGURE 17.3   O-type water tube boiler furnace area. Sunmarks, LLC.

boiler or DA could be an indicator of fouling and oxygen corrosion. Unstable boiler water level maybe an indication of poor boiler water chemistry. DA water temperature should correspond to the pressure being maintained in the vessel. If the two do not correspond well, the DA might be stratified or the deaeration section may be damaged. The burner flame color and pattern should be observed routinely to ensure good combustion is occurring. A gas-fired burner should have a nice blue flame at all firing levels. If the flame is yellow or orange, there may be too much excess air being introduced to the burner.

Monthly inspections are more related to material condition and performance. One inspection often overlooked is the boiler stack temperature. This single inspection data point can tell a lot about how clean the boiler pressure vessel is. A scaled or fouled pressure vessel will allow for less heat transfer and account for higher stack temperatures. Generally a boiler will not foul or scale acutely. Therefore trending of the boiler stack temperature is required to fully predict what is happening inside the unit. Remember the stack data will only be useful if the temperature is recorded at a constant boiler firing rate. Periodically manually setting the firing rate at some level for a short period of time then record the stack temperature is a good practice.

Other monthly inspection areas are feed water pump motor amp checks and pump discharge pressure. Pump internal and bearing wear can usually be predicted if the pump performance is trended. Boiler and DA gasket integrity for manways and handholes, chemical feed pumps, chemical feed lines, and pressure and temperature gauges should be check. Steam traps should also be monitored on a monthly basis. In a process application where it generally is noisy, the best way to monitor steam traps is to use a temperature sensing device downstream of the trap to monitor condensate temperatures. If the condensate temperature starts to increase significantly, the trap may need to be repaired or replaced. The burner/boiler shell plate temperature should be monitored at least monthly to ensure the burner refractory is intact. A hot spot may be an indication that the refractory needs repair. Another good monthly check is to look at the fuel usage per steam production. Essentially this is the fuel to steam efficiency. This can be done if the steam and fuel are metered continuously. Simply take a few data points and record fuel input and steam output. The ratio should be relatively constant.

If the efficiency starts to decrease, the boiler might be scaled or fouled or the steam traps might be leaking by. A monthly inspection guidance is included in Appendix B.

System shutdown inspections. When the steam system is removed from service, internal inspections of the boiler pressure vessel and feed water de-aerator should be accomplished. The main purpose of these inspections is to verify the water treatment scheme is working to protect the equipment. Once the boiler and DA have cooled and have drained, open the manways and perform a visual inspection. It is highly recommended this inspection is performed with the water treatment representative there to witness to internal condition with you. The internal vessels should be clean and the metal a very blackish color; this blackish color is indicative of a good magnetite layer being formed on the metal surface. If the internal metal shows a reddish-orange color, then you likely have oxygen-related corrosion. The images in Figure 17.5 show the color differences between the two types of corrosion.

The inlet flange/threaded connections for the safety relief valves and the valves themselves should also be inspected minimum once per year. The reason for this is to predict SRV performance in an emergency condition. The inlet and SRV valve internals should be free of debris. The boiler furnace area should also be thoroughly inspected at least annually. You should look for any damage to the burner and refractory. Damage to either should be repaired as soon as possible. The DA spray nozzles or trays should be inspected for cracks or wear. It is a good practice to take pictures of the vessel internals and put them in an equipment file. Comparisons of these pictures will help you to verify if your preventative maintenance program is working (Figure 17.4). An inspection form is included in Appendix B.

BOIL OUT AND LAYUP PRACTICES

After installation of a boiler and feed water system they should be thoroughly rinsed prior to being placed in service. This cleaning process is often called a "boil out." A good boil out procedure will remove oils, greases, pipe thread compound, solvents, and any other contaminants in contact with the water side of the pressure vessel. These contaminates can cause pH excursions and foaming if not removed. A complete

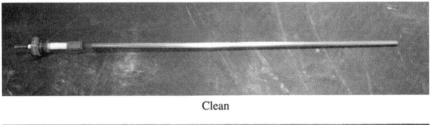

Clean

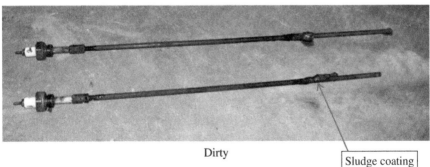

Dirty

Sludge coating

FIGURE 17.4   Clean and dirty water level probes.

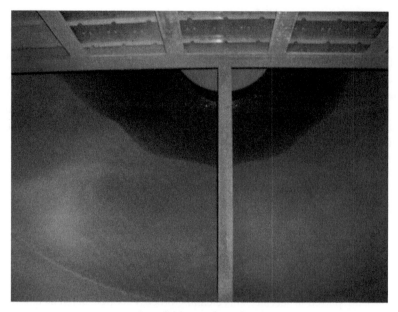

Reddish iron oxide—bad

Magnetite (black) iron oxide—good

FIGURE 17.5   Carbon steel pressure vessel iron corrosion.

boiler out procedure is included in Appendix B. Essentially, a surfactant-type chemical is added to the feed water tank and pumped to the boiler where it will be heated up and dissolve any unwanted contaminates. The contaminated water is completely drained and the system is refilled.

If the steam system is to lay idle for a period of time (3 days or longer) then it is a good practice to "layup" the system properly. Layup procedures can be either a **wet layup or dry layup**. The wet layup procedure is used when the system is taken out of service for a short period of time, like a few weeks to a month. A dry layup procedure is used for a steam system being taken out of service for more than a couple months. The primary goal of each of the layup procedures is to prevent excessive corrosion during the idle period. Specific guidance for each layup procedure is included in Appendix B.

Wet layup will keep the feed water and boiler water in their respective vessels and use a dose of oxygen scavenger in each vessel to keep the dissolved oxygen levels at a very low level. Keeping the dissolved oxygen levels at a low level will effectively stall any general corrosion. Using a chemical treatment to remove oxygen is not an effective method over a long period. Consequently, for these prolonged idle periods require the system to be drained and dried out. A dry layup procedure will allow the system operator to remove all the water from the system, open the vessels, and install a heat source to keep the internal metal dry. Many times a simple incandescent light is put in the boiler and feed water tank to add just enough heat to keep the metal surface dry. A dry metal surface will not corrode.

18

TROUBLESHOOTING AND COMMISSIONING BASICS

This section of the book will provide the commissioning engineer, service technician, or operator a useful approach methodology and strategy to satisfactorily start up, commission, and operate a process steam system. Guidance for precommissioning and troubleshooting techniques is presented. An efficient startup and commissioning will optimize total system performance in addition to individual equipment performance.

Startup Versus Commissioning

Commissioning differs from startup. Startup verifies each piece of equipment in the system is performing its design function. Commissioning verifies the entire set of equipment is working in unison to perform its intended function. Think of startup and commissioning as similar to the functions of a NASCAR driver and the pit crew. The pit crew services the vehicle, verifies the car runs correctly, and is adjusted right for the driver. The pit crew therefore provides the startup function. The driver, however, is the commissioning engineer. This individual drives the car and makes sure the car can perform its function on the track, operating at all speeds and track conditions.

Consequently, the driver needs to listen and feel the car as they run the race and tell the pit crew what they think can be done to improve the car's performance. Likewise, the commissioning engineer must also be able to listen, observe, and determine if a steam system is operating correctly and, advise what adjustments can be made to allow the system to operate at peak performance. The commissioning engineer must be a good listener and communicator to be effective.

Process Steam Systems: A Practical Guide for Operators, Maintainers, Designers, and Educators, Second Edition. Carey Merritt.
© 2023 John Wiley & Sons, Inc. Published 2023 by John Wiley & Sons, Inc.

Approach to Troubleshooting

In a perfect world, the maintenance and operations staff work closely with the design engineers, installing contractors and the equipment manufactures to deliver a good performing system. Rarely, however, will the design and installation alone be perfect. Therefore, startup technicians and commissioning engineers may need to troubleshoot and adjust the equipment to achieve high performance. Good troubleshooting skills are essential and include

1. Documentation of "as found" conditions of the system. This needs to include data from talking to the operators, installing contractors and anyone who has recently touched any of the system components. Material condition, tank levels, valve positions, pump performance data, heat exchanger performance data, chemical feed rates, history of failures, peculiar smells and sounds. etc. all should be written down prior to making any adjustments. Do not forget to document any other system interface problems (i.e., electrical breaker issues, supply water and gas pressure, etc.). Much information can be obtained from talking to the operators and maintenance folks. The precommissioning checklist found in Appendix C can be used to record as found conditions. It is very frustrating for an owner to read a service or commissioning report that does not include detailed "as found" conditions.

2. Take time to understand the design basis. We all have paradigms that make us try to fit all new system operation into some other system you are familiar with. I have heard "the last system I worked on did not act like this" many times from service technicians. You must approach all new systems with an open mind and allow it to operate as it was designed. Verify supporting systems are supplying adequate support. This is especially important for combustion air and exhaust gas stacking. Determine if room air pressure is adequately being maintained. Is a combustion air unit operation logic tied to boiler fan operation?

3. Systematically eliminate good functioning equipment or subsystems within a piece of equipment as a source of a problem. For instance, if a boiler does not run properly, then eliminate the subsystems like level controls, gas pressure, stack pressure, etc. that do function one at a time.

4. Change only one variable at a time. Failure to adhere to this will cause frustration. We all want to jump to conclusions and adjust something that we know fixed a similar problem in the past. Be very careful not to adjust more than one parameter at a time. Good troubleshooting has to eliminate causes one step at a time.

5. Keep good notes. Record every adjustment, time, symptom, fix, conversation, success, and failure you have. Sixth months after you think you are done and get a call from an anxious owner, your notes will become very valuable.

Don't Play the Blame Game

Care should be taken not to counter advice or recommendations supplied by the equipment manufacture unless their guidance will create a safety hazard. Occasionally, a system design concept may have been altered from its original intent to meet budget requirements. For instance, the condensate system installed for a particular system wastes the condensate to drain and that causes excessive make up water use and poor feed water quality. You may find that the original condensate system design showed all the condensate returning to the feed water deaerator; however, during installation the cost of running the condensate line back to the deaerator was excessive and the installer deleted this work from the scope of the project. It is never appropriate for the commissioning engineer to determine blame for a deviation in design or installation. Owners may ask you to do this; however, you should report facts and performance only and stay away from the blame game. The same philosophy should be used to document installation issues that pertain to safety concerns or equipment accessibility. All discrepancies from the design to installation to the method of operating the equipment need to be documented clearly with the emphasis on system performance and reliability. The commissioning engineer is the liaison between the design engineer, equipment manufacturers, salesman, installer, and the owner. Make absolutely sure that any recommendation you make does not counter any information contained in any of the equipment O&M manuals.

Compliance with O&M manuals can be used to determine liability for a system problem in a court of law. It is appropriate, however, to notify the owner when information in an O&M manual is unclear or conflicts with standard practices.

Precommissioning

Prior to commissioning a new or retrofitted system, the engineer must obtain some basic data sheets and ensure that the steam system precommissioning checklist shown in Appendix C is completed. Precommissioning verifies installation and equipment startup activities are complete. Many system commissioning efforts have failed because the precommissioning activities were not correctly completed. You would not attempt to drive across the country unless your vehicle was ready for the trip. The same analogy can be used for commissioning a steam system. The following information should be collected prior to commissioning.

1. Specification used to design the system, including any changes
2. Copies of the submittal drawings including P&ID's and electrical logic drawings
3. Copies of all factory test sheets…, that is, test fire, hydro test, ASME P2 data sheets, etc.
4. Copies of startup and any pertinent logs
5. Any maintenance records including water treatment schedules
6. All change orders from the installing contractors
7. Component cut sheets…, that is, pump curves, valve and regulator cut sheets, etc.
8. Records of any shipping and installation issues
9. O&M manuals for all major components

In addition, the following tools may be necessary for the commissioning engineer to verify installation specifications and system performance.

1. Flashlight
2. Measuring tape
3. Small level
4. Notebook
5. Stack gas analyzer
6. Calculator
7. Stopwatch
8. IR temperature gun
9. Voltmeter or test light
10. Camera

19

COMMISSIONING AND TROUBLESHOOTING THE STEAM GENERATOR

The steam generator, like the ones shown in Figure 19.1, is the heart of the steam system. It is the piece of equipment that converts fuel energy into steam. Steam generators or boilers can range in size from very small (30 lb/h) to very large (>100,000 lb/h) The commissioning engineer should determine

1. Is the steam generator installed correctly?
2. Is it large enough to meet the steam load demand?
3. Is the turndown lower enough to prevent excess burner cycling?
4. Is it trimmed in a manner to ensure ease of operation?
5. Can the steam generator be operated and maintained to ensure reliability?

Before these questions can be answered, you must first determine actual input, output, and efficiency of the steam generator. In addition, you must evaluate the water level, steam and fuel pressure, and burner and blowdown controls and ensure all function well enough to support sustained steam operations. **Steam Generator Commission Data Sheet** in Appendix C can be used to document this evaluation and should be periodically reviewed when reading this section.

Determining Boiler Input, Output, and Efficiency

All boiler manufactures list the output of their boilers relative to 0 psig steam pressure and 212 F feed water temperature. Under these ideal conditions all of the fuel energy transferred to the water (minus boiler radiant losses) will be used to overcome the

Process Steam Systems: A Practical Guide for Operators, Maintainers, Designers, and Educators, Second Edition. Carey Merritt.
© 2023 John Wiley & Sons, Inc. Published 2023 by John Wiley & Sons, Inc.

FIGURE 19.1   The steam generator. Hurst Boiler and Welding Co.

latent heat of vaporization to create steam. We all know that a boiler will never operate under these conditions, so we must calculate actual output. The first step in determining output is to accurately determine input.

The best way to document input is to simply read a fuel meter inline to the boiler burner. Sometimes, a meter is not installed and input must be verified by adjusting fuel pressure in the field to match the fuel pressure used at test fire at the factory. One can generally assume if the burner inlet fuel pressure in the field is similar to what is found on the factory test fire sheet, then the fuel input will be the same as documented by factory personnel. This verification should be done at high and low fire operation. A fuel pressure gauge on the last elbow (gas fired) or the fuel pressure to the burner nozzle (oil fired) prior to the burner head should be used to document fuel input. Generally, the gas pressure should be within ±0.5 inches water column pressure and oil pressure should be within ±5 psig at the burner inlet relative to the factory test fire sheets.

Calculating actual output can be a little tricky. Exact output can only be measured in a lab where all losses can be accounted for and input is precisely measured. Since the lab conditions do not exist in the field, a commissioning engineer must realize that the output calculated in the field will have some inherent error. Realistically, if you can measure output to ±10% of the projected output, you are doing well. First, I recommend obtaining from the manufacture, the projected output based on installed conditions. Actual stack draft, fuel btu content, fuel pressure, and boiler operating pressure will be required by the manufacture technical staff to project the actual output. Asking the manufacture for the expected fuel to steam efficiency of the boiler at the installed operating conditions is recommended; however, you might only get the expected output in btu.

Once the input is known and the manufacture provides the efficiency, the output can be calculated by multiplying the input by the fuel to steam efficiency and dividing by the amount of energy required to make 1 pound of steam from 1 pound of feed water at the boiler operating parameters (i.e., feed water temperature and steam pressure). The formula is shown below.

$$\text{Input fuel energy} (btu/h) \times \text{fuel to steam efficiency} (\%)$$
$$= \text{Available energy to make steam} (\mathbf{Qs})$$

And $\mathbf{Qs}$ in $btu/h/(\text{steam temp} - \text{feed water temp} \times 1 btu/lb) + \text{latent heat} (btu)$
$$= \mathbf{lbs/h\ steam\ output}.$$

Example 19.1

100 hp steam boiler operating at 90 psig at high fire is expected to have a fuel to steam efficiency of 80%. The manufacture indicates the installed boiler will have an input of 4,000,000 btu/h with the gas pressure seen at the installed boiler burner. The system is designed to have 50% condensate return. Makeup water temperature is 70 F. What is the projected output?

Answer: Qs = 4,000,000 btu/h × 80% = 3,200,000 btu/h available energy for steam production. And the feed water temperature with 50% returns at 212 F will be 212 F − 70 F/2 = 140 F. The actual steam output from this boiler should be

$$\textbf{3,200,000btu / h / (331F} - \textbf{140F} \times \textbf{1btu / h)} + \textbf{887btu / lb} = \textbf{2968\ lbs / h}$$

Note: 331 F is the saturation temp of steam at 90 psig, 887 btu/lb is the latent heat of vaporization at 90 psig, and 70 F is the cold water makeup temperature. If the fuel to steam efficiency is not known, then it can be calculated by subtracting the radiant and blowdown heat losses from the published output.

Boiler Performance Test

Now that you have calculated the boiler output at the installed conditions, you can verify this with a simple but effective method of measuring steam output. Please refer to Figure 19.2 to see the general guidelines for performing the test procedure. The steam output of a boiler can be directly related to the water used by the boiler. If you do not discharge boiler water (i.e., blowdown) during a specified period and the boiler steam is wasted to atmosphere, the steam output can be determined by measuring the total amount of makeup water added to the boiler over a given period of time. Actual output should be within ±10% of the calculated output.

Performance test procedure: See Appendix C for specific procedure. Ensure the load on the boiler will not change significantly over the next hour and install a water meter in the makeup water line to the feed water/DA tank. Bring the boiler up to operating pressure and establish steady state (you may have to dump steam to achieve

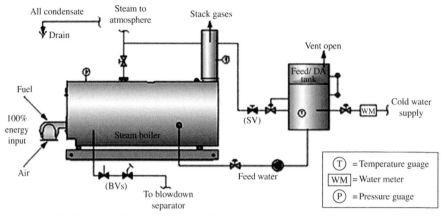

Simple steam boiler test setup to measure fuel to steam output

Note : 1. Blowdown valves (BVs) and steam preheat valve (SV) must be closed during the test.

2. Minimum combustion efficiency should be start prior to test.

3. Steam load and steam pressure must be constant during test.

4. Test starting and stopping water level in the feed tank/ DA should be the same.

Steam out = water usage

FIGURE 19.2   The boiler output performance field test.

steady state). You should isolate the steam supply to the feed water tank/DA and allow the DA to reach steady-state temperature prior to testing the boiler output. Any condensate should be wasted during the test period. Record boiler test start time and water meter starting gallon reading. Let the boiler operate for at least 1 h under these steady-state conditions and then record the water meter reading, and the actual feed water temperature. Subtracting the finish water meter reading from the starting reading and dividing by the test period in hours will give you a fairly accurate measure of actual steam output in lb/h. It is best to start and stop the test when the level in the feed water tank is the same level at the test beginning and end.

Example 19.2

*In the boiler mentioned above in Example 16.1, you ran the boiler at 90 psig for 90 min and used 490 gallons of feed water. The boiler ran at high fire and the feed water temperature was 100 F. If the manufacture states the boiler output is 3,200,00 btu/h, what is the actual output? How does it compare to the calculated output? The actual output is 490 gallons × 8.34 lb. Gallon/1.5 h = 2724 lb/h steam. The calculated output would be 3,200,000 btu/h/(331 F − 100 F) + 886 btu = 2865 lb/h. A comparison of actual versus calculated output is then 2724/2865 = **95.1% of projected output**.*

Note: If the measured output > calculated output, then you likely have excessive carryover of boiler water. In this case the header drip legs can be drained to a container during the test and the amount of water collected can be subtracted from the water meter readings to give a more accurate test result.

Example 19.3

*The test performed above on the same boiler showed the steam output to be 4500 lb/1.5 h. This equates to 3000 lb/h which is greater than the calculated 2865 lb/h. Draining a drip leg from the main steam header to a 55 gallon drum showed about 34 gallons/h carryover. Subtracting this carryover water, the actual output of the boiler would then be 3000 lb/h – (34 gph × 8.34 lb/gal = 300 lb/h) = 2700 lb/h. Note that 2700 lb/h actual output **94.2%** of the 2865 lb/h calculated output.*

In most cases, the installed boiler (s) will have been sized for maximum load projections and will be oversized. The same test can be repeated to determine boiler minimum output provided the boiler burner is locked on low fire during the test period.

Commissioning the Boiler Burner Controls

The burners function is to add fuel and air together, and combust them in a safe efficient manner. When the boiler calls for heat (the steam pressure falls below the set point), the burner pilot lights, flame is established, and within a few seconds the main flame ignites on a low fire setting. The burner firing rate will then drive toward high fire and modulate at a firing rate that will produce just enough steam to satisfy the demand.

Commissioning the burner controls can be done by observing the flame programmer, burner air, and fuel supply valves during the prepurge, pilot lighting, main flame lighting, and the burner shutdown operations including postpurge cycles. The flame programmer display will show the status of various phases on the burner light off cycle. The commissioning engineer should verify the adequacy of the pre- and postpurge and ensure the burners response to load changes satisfies the application.

Commissioning the Boiler Pressure Control System

Commissioning the pressure control system is simply verifying the operating and high limit pressure controls are set correctly for the application. This part of the steam system controls the rate of steam generation. Since the boiler does not measure mass flow of steam, it uses steam pressure as a means to regulate output. Generally there are at least two pressure sensors on the boiler located somewhere in the steam space connection that control the boilers operating pressure and provide a high-pressure safety interlock. The operating pressure control may be on/off switch type or modulating operating pressure control type. The high limit is always an on/off switch and will require a local manual reset. The commissioning engineer must observe the boiler operation through a few cycles and make adjustment recommendations that will keep the steam pressure required at the steam-using equipment without causing excessive boiler burner cycling.

Commissioning the Boiler Level Control System

The level controls have two distinct purposes. One, they control operating water level in the boiler, and two, they provide a safety shutdown interlock in case of a low water condition. Level control problems can create an unsafe condition, poor performance, or nuisance shutdowns. The commissioning engineer needs to verify the level controls adequately keep the water level in the boiler in a safe condition and will shut down the boiler if the water drops to an unsafe condition. There are two types of water level controls: on/off and modulating.

To commission the level control system, observe the level over a 30 min time during a period where the boiler is operated at or near the rated pressure and steam flow. The water level should be maintained visible in the site glass between the normal level high and low marks during this period. It may warrant using a marker pen to put marks on the boiler sight glass to show these various levels.

The high water alarm can be checked, when the burner is off, by placing the feed water pump in "hand" and manually overfill the boiler until the high water alarm activates. The water level can be drained slowly until the first low water cutoff level is achieved. A burner lockout signal should be observed. Refill the boiler and verify the burner lockout will reset. Next redrain the boiler until the second low water cutoff level is achieved. This level should lock out the burner and require a manual reset at the boiler.

The importance of this part of the boiler commissioning cannot be overemphasized. Low water boiler explosions have killed or injured thousands of people over the last century and a half. Fortunately we have codes that have tremendously reduced this potential hazard. If you find the low water circuitry tampered with or altered from the factory, you should immediately notify the owner and have the level control system integrity restored prior to completing more work on the boiler.

Commissioning the Boiler Blowdown Controls

To avoid boiler problems, water and solids must be periodically discharged or "blown down" from the boiler. The commissioning engineer must ensure the blowdown scheme and procedures keep the boiler water chemistry within specification set forth by the boiler manufacture. Boiler logs and water chemistry reports can help determine the effectiveness of the blowdown practices.

Steam Boiler Troubleshooting

Problem: Boiler burner does not come on when the boiler on/off switch is placed in the "on" position.

Cause: The flame programmer may be faulty, the boiler water level may be outside the normal band, the steam pressure may be higher than the set point, or the power supply to the boiler panel may be off.

Consequence: No steam output.

Adjustment: Verify power is "on" to the panel box, the water level is at a normal level in the sight glass, and the steam pressure is below the set point. With these conditions are met, then the flame programmer should display the "prepurge cycle." If the flame programmer does not indicate any reading, then power to the programmer is absent or the programmer is not functional. Troubleshoot/replace as necessary.

Problem: Boiler combustion air fan starts but the burner does not light.

Cause: Air switch is not proving, the prepurge end switches are not proving, or one of the conditions outlined in the above problem are not met, that is, water level or steam pressure.

Consequence: No steam output.

Adjustments: Place **temporary** jumper on air switch circuit to verify air switch is functional. Verify prepurge end switches are making contact. Adjust/replace as necessary.

Problem: Boiler frequently shuts down on flame failure.

Cause: Flame failure can be a result of poor combustion, poor stack draft, inadequate fuel supply, or poor flame sensing. All of these will need to be right for consistent light offs and good combustion.

Consequence: Excessive fuel use, burner furnace sooting, main flame failure shutdowns, excessive burner noise, or no steam output.

Adjustments: Combustion is usually set at the factory but may be reset at different conditions that are found in the field. Obtain factory test fire sheet and make sure fuel, air, and draft conditions meet the requirements for that boiler. Using a gas analyzer, check stack CO_2, O_2, and CO levels to make sure they are consistent with the factory test fire sheet. High stack O_2 levels mean high excess air. High stack CO levels may mean lean combustion. Stack gas analysis should be performed at low, mid, and high fire for all fuels being combusted. If combustion appears to be okay, then check the flame sensors. If flame sensor is faulty, the burner will not maintain pilot and/or main flame.

If the sensor orientation is poor, the sensor may produce inconsistent flame sensing. Replace or reorient the flame sensing device.

Problem: Boiler burner cycles a lot.

Cause: The burner/boiler is oversized; the pressure controls settings are too close or the steam chest is too small.

Consequence: Premature burner failure and loss of efficiency.

Adjustments: First, ensure the pressure switch pressure settings, that is, set points and differentials are set at the maximum levels to allow the burner to stay on over a wide range of system pressure. The process equipment being served by the boiler will limit this range. On/off pressure switches can be replaced with PID (pressure integral differential)-type pressure controls. PID-type controls give additional flexibility for burner response to system pressure. The burner may be able to be derated to lower the output.

A steam accumulator can be added between the boiler and load to increase the steam storage capacity.

Problem: Boiler burner tends to "howl."

Cause: When the draft is too negative, it will pull the flame off the burner head. If the draft is way negative the flame will go out; however, if the draft is slightly more negative than recommended, the flame will actually collapse back onto the burner head. It is this constant pulling and collapsing that creates a high-pitched noise called "howling."

Consequence: Nuisance flame failure shutdowns and burner noise.

Adjustments: Install barometric damper in the stacking to reduce high negative draft. If an exhaust fan is installed in the stacking, then adjust pressure/fan speed to reduce negative draft condition. A stack economizer can be added that will increase stack back pressure.

Problem: Burner never comes off low fire.

Cause: Pressure controls are not allowing the burner to reach high fire or, the modulating linkage is not adjusted properly.

Consequence: Boiler output will be limited to the low fire output level.

Adjustments: Disconnect linkage from damper/fuel valve and verify full motion of modulation motor or servos during the prepurge cycle. Replace faulty mod/servo motors or limit switches. If the pressure controls are via a PID-type pressure controller, then ensure the "P" or proportional band is set properly. If "P" is set too high, then the controller will not provide a signal that allows the burner to reach high fire.

Problem: Boiler burner flame has bright orange color.

Cause: The air to fuel ratio is too high and the burner is over aired. This is found frequently when a burner turndown level is higher than a level the burner will support good combustion.

Consequence: Combustion efficiency is significantly reduced. Sooting may very well occur if the burner is operated under these conditions for a prolonged period.

Adjustments: Adjust air gate to reduce air until good combustion occurs. The flame should be relatively blue throughout the firing range.

Problem: Boiler stack analysis shows high carbon monoxide levels.

Cause: Burner is being operated too lean. CO and O_2 levels are inversely related in a stack gas analysis. CO levels should be <300 ppm.

Consequence: Poor efficiency and potential health hazard.

Adjustments: Open up the air supply damper or reduce the gas flow to the burner. Combustion air intake temperature compensation or stack oxygen trim can be added/adjusted to maintain CO levels <300 ppm.

Problem: Boiler does not reach full pressure.

Cause: The boiler is undersized. Any steam boiler will try to satisfy the steam demand. If the demand cannot be satisfied, the boiler will not be able to supply enough steam in the piping to build pressure.

Consequence: The load equipment will be starved of steam, the lower pressure steam will have a lower temperature, and the steam supplied will likely have more entrained water.

Adjustments: Increase the output of the existing boiler, shed unnecessary steam loads, or add more steam generation equipment.

Problem: Boiler pressure drop rapidly and returns to rated pressure relatively quickly.

Cause: Instantaneous demand for steam exceeds the steam storage capacity of the boiler and piping. Examples: A steam control valve opening too fast, supplying steam to a cold metal vessel, or supplying steam to an evacuated autoclave.

Consequence: Rapid depressurization of the boiler will cause boiler level expansion and poor steam quality and create a condition that favors low water shutdowns.

Adjustments: Slow down the control valve opening rate, close down the boiler stop valve, or install an orifice plate in the boiler steam outlet piping. Adding a wet or dry steam accumulator will increase the system steam storage.

Problem: Boiler pressure drops rapidly and stays low for an extended period of time.

Cause: Boiler output is too low. See boiler does not reach pressure problem.

Consequence: Prolonged cycle times.

Adjustments: A steam accumulator will help but not solve the problem. You will need a bigger boiler or you will need to shed some of the existing steam load.

Problem: Boiler shuts down on high pressure requiring a manual reset.

Cause: The operating pressure control limit is too close to the high-pressure limit switch. This problem is often associated with steam boilers that are set up to operate very close the high limit pressure setting. Once a burner shuts off, there is still enough residual heat to allow the pressure to creep up past the operating limit a few psig. This boiler trip will create a manual reset condition.

Consequence: Nuisance shutdowns.

Adjustments: Widen the gap between operating and high limit set points.

Problem: Boiler frequently shuts down on low water.

Cause: 1. Boiler level control system circuitry may be malfunctioning. This may be a level sensor, water level relay, pump motor, or wiring between each component.

Consequence: Boiler out of service.

Adjustment 1: Take continuity measurements from the level sensor to each component and determine where the failure is located. You may have to cycle the level in the boiler to see what component does not respond to the level control signal. Replace failed component.

Cause 2. The feed water pump may not be large enough.

Consequence: Nuisance low water shutdowns. Typically an undersized pump will not be able to meet flow or discharge pressure requirements. Trying to pump very hot water with not enough static head will cause a pump to cavitate (sounds like marbles are in the pump).

Adjustment 2: Obtain pump curve and verify flow rate at the discharge pressure observed and required head pressure (NPSHR) of the pump. If any of the three do not meet the system requirements and cannot be throttle to meet the requirements, then a new pump is required. Also check to see the pump inlet strainer is not plugged, preventing flow.

Cause 3. The feed water pump may not be running on the desired part of the pump curve. Too low a discharge pressure may cause pump run out and NPSHR concerns, and too high a discharge pressure may cause too low of a flow rate to the boiler.

Consequence: Nuisance boiler low water shutdowns.

Adjustment 3: See above fix. If condensate is returning at the boiling point (IR gun temp on the feed water tank), then leaking steam traps may be a problem. Normally, if pump run out is experienced, then throttling the discharge valve on the pump will align the pump output with the desired rate on the pump curve.

Cause 4. The boiler may be experiencing quenching caused by injecting relatively cold water into a hot boiler and collapsing of the water column.

Consequence: The hot expanded water in the boiler suddenly is cooled by adding relatively cold feed water and the water is cooled rapidly enough to collapse the steaming rate and abruptly lower the boiler water level. The feed water is not hot enough and/or it is being added at a very high rate.

Adjustment 4: Ensure feed water is at least 200 F. Throttle back on the feed water pump discharge valve to slow feed water injection rate. If boiler is equipped with on/ off feed water controls, you can change the feed water controls to a modulating type.

Problem: **Idle boiler floods**.

Cause: Lack of high water protection in the boiler. Tramp steam/water from the operating boiler is condensing in the idle boiler, or the idle boiler pulls water into the boiler when the boiler cools down and contracts.

Consequence: Requirement to drain water from the boiler, poor steam quality and wasted water.

Adjustment: Ensure high water protection is installed in each boiler. This may be an equalization line connecting the boiler water volumes, or a high water trap assembly installed about 1–2 inches above the normal boiler water line. Tramp steam may also be captured before it enters the idle boiler by adding a check valve/nonreturn valve (high-pressure systems only) or a drip leg assembly on the steam header downstream of each boiler. Each boiler should also be fitted with a vacuum breaker to prevent pulling water into the idle boiler when the boiler cools down.

Problem: **Erratic water level oscillations in the boiler**.

Cause: Water level oscillations are symptom of poor pressure control, poor level control, and/or poor boiler water chemistry.

Consequence: Nuisance low water shutdowns and poor steam quality.

Adjustment: Ensure steam control valves downstream of the boiler are not opening and closing too fast. Boiler total dissolved solids (TDS), and alkalinity levels increase water surface tension and need to be kept within limits specified by the boiler manufacture. Also any tramp organics like solvents or oils in the boiler water will create water level bouncing. For all these conditions, completely drain and refill the boiler. If problem persists, then reperform a boil out of the boiler. Also, you may be able to add a short time delay relay to the low water cutout circuit to mitigate nuisance low water tripping caused by dips in water level.

Problem: Boiler does not shut down on low water conditions.

Cause: Faulty low water cut off circuitry. This condition should never occur with a new boiler, but the consequences are severe enough that I mention it here anyways.

Consequences: Boilers operated without enough water can explode and/or cause severe damage to the pressure vessel integrity. Never add water to a boiler you think may be in this condition. Turn the boiler off and walk away until it cools down.

Adjustment: The fault may be in the sensing device, wiring, or the relay designed to shut the burner off in the event of a low water condition. Using an ohmmeter, start with the relay and work toward the sensing device to determine where the fault is located. This testing should be done when the boiler is energized but not operating.

Problem: Boiler experiences high carryover.

Cause: Poor steam quality is caused by high steam exit velocity, high boiler water levels, or improper water chemistry.

Consequence: The heat transfer rate is reduced when water is present in the steam. Contamination of the steam-using equipment may occur if excessive carryover is not adequately removed from the steam headers.

Adjustment: The solution to a carryover problem can be corrective or preventative. Corrective measures include adding steam separators or steam filters in the steam lines to remove the entrained impurities. Restricting orifice plates can be installed in the steam line near the boiler outlet to help mitigate the consequences of a rapid boiler depressurization event. Preventative measures include lowering the boiler water level slightly, better water chemistry maintenance, adding an accumulator or even increasing the boiler operating pressure.

Problem: Boiler internal inspection or boiler water shows rust colored deposits.

Cause: Excessive iron corrosion is occurring somewhere in the steam system. Sometimes the iron corrosion products (responsible for the reddish-orange color of the water or deposits) collect in the boiler but may originate from corrosion of the steam lines, condensate line, or feed water equipment. This is an indication of accelerated general corrosion likely due to low pH and/of high dissolve oxygen levels somewhere in the steam cycle.

Consequence: The high iron levels indicate the boiler pressure vessel, feed water tank, or piping is corroding at a high rate and premature failure of some of the steam cycle equipment may occur. If iron levels become high enough, iron may precipitate out on heated surfaces reducing heat transfer and boiler efficiency.

Adjustment: Start a sampling scheme that measures insoluble and soluble iron in various places in the steam cycle. Steam drip legs, condensate lines, feed water tank, and boiler samples should be analyzed to see where the source of iron corrosion is occurring. A good place to start is the boiler and work forward.

Problem: Boiler experiences sudden high stack temperature.

Cause: When the stack temperature rises in the boiler exhaust, this is a result of a loss of thermal efficiency in the boiler. The sudden loss of efficiency is likely due to a problem inside the boiler furnace. A broken refractory or baffle could cause the flame to pass through the boiler furnace too quick to facilitate good heat transfer. A sudden negative change in draft venting can cause the high negative draft to pull the flame through the boiler and cause stack temperatures to spike.

Consequence: Poor overall boiler efficiency.

Adjustment: Verify stack draft conditions are within spec. An internal inspection of the boiler furnace area will be required and repair of any broken refractory or baffles will need to be accomplished.

Problem: Boiler experiences gradual increase in stack temperatures.

Cause: Boiler waterside fouling or scaling. Severe fireside sooting will also cause a high stack temperature.

Consequence: Loss of boiler efficiency.

Adjustment: Shut down the boiler and clean furnace area and internal waterside of the boiler pressure vessel. If there is hard scale formed on the boiler internals, then removal may be difficult.

Problem: Boiler burner experiences severe vibrations.

Cause: Inadequate back pressure in the furnace area.

Consequence: Nuisance noise and vibrations, burner damage.

Adjustment: A burner adjustment by the burner manufacture should be performed. Worse case you may have to install a blocking wall in the pressure vessel furnace area to stabilize the back pressure on the burner.

20

COMMISSIONING AND TROUBLESHOOTING THE STEAM DELIVERY SYSTEM

The steam delivery system is the essential link between the steam generator and the steam user. An efficient steam delivery system is critical if steam of the right quality and pressure is to be supplied, in the right quantity, to the steam-using equipment. Commissioning of the steam delivery system verifies the design and installation of the components is correct and will perform its intended function. The steam delivery system is comprised of the steam distribution piping, control valves, condensate draining, and venting equipment and specialized equipment. The **Steam Delivery System Commissioning Data Sheet** in Appendix C can be used to document system adequacy.

Steam Distribution Piping

The steam delivery system starts with the steam distribution piping, similar to what is shown in Figure 20.1. Correct piping size and orientation are essential for proper operation of the entire steam delivery system. There are a few rules of thumb that should be field verified for the steam distribution system.

- The piping should be sized for steam velocities 6000–12,000 ft/min, the lower for steam heating applications and the high for process applications.
- The piping should have a slight slope gradient toward the flow of steam.
- The drip legs should be sized at ½ the diameter of the main header diameter (but equal in size to header if header is <4 inches) not less than 2 inches and 18" minimum drop.

Process Steam Systems: A Practical Guide for Operators, Maintainers, Designers, and Educators, Second Edition. Carey Merritt.
© 2023 John Wiley & Sons, Inc. Published 2023 by John Wiley & Sons, Inc.

FIGURE 20.1   Undersized and poorly drained steam header.

- The boiler supply steam line should always come into the top of the main steam header.
- Restricting orifice plates can be used to retard the flow of steam during high demand startup loads. This is a useful tool for low-pressure design.
- An air vent/vacuum breaker should be placed on the steam main header to prevent condensate aspiration back into the boiler upon cool down and to allow air to be vented out and replaced with steam.
- The main steam piping should always be insulated.

Control Valves

Control valve selection is as important as pipe sizing. These valves control steam flow and pressure to maintain a desired temperature of the product being heated. Since temperature and steam pressure are related, simply controlling the steam pressure in the steam lines to the product will regulate the amount of energy available for the product to absorb. System commissioning engineers must ensure slow opening times or rapid boiler depressurization can occur. When a steam valve slams open quickly, the steam pressure falls in seconds and boiling in the boiler become erratic due to rapid bubble size increase. Opening of control valves too fast will cause carryover and rapid level drop in the boiler. Pneumatic control valves are subject to rapid response and must be adjusted to regulate valve stroke times. Valves should be verified to be the right size and provide the level of control as needed. Control valves should have no more than 2–4 psig pressure drop when they are fully open.

Steam Piping Venting

Venting is also very important function of the steam delivery system. Venting is simply providing a pathway to remove air from the system during startup and allowance for air to reenter after the system is shut down. Typically placing thermostatic air vents in the steam piping will perform both functions adequately. It is important to remove air from the system during startup to prevent "air binding." Likewise, when they are allowed to cool down, they reopen and allow air to flow again. They act as a vent and a vacuum breaker. Commissioning engineers need to ensure the steam delivery system has the appropriate means of venting installed.

Condensate Trapping/Draining

Condensate starts to form as soon as the steam leaves the boiler. Consequently, steam headers need to have drip legs to allow the condensate to drain from the steam lines. Steam traps are used to provide a barrier that will separate the condensate from the steam. Steam traps come in a variety of sizes and functionality. Applying the correct size and type of trap is key to good condensate drain system design. Incorrect application can lead to system heat transfer inefficiencies and wasted money. Likewise trap location relative to the heat exchange equipment is essential to ensuring proper condensate draining. The right trap selection and size should be verified. In addition, the condensate piping should be verified to be sized and oriented correctly to facilitate good draining. Condensate return lines entering a main return header should have diffusers installed to prevent steam hammer.

Troubleshooting the Steam Delivery System

Problem: The steam lines are producing a whistling noise.

Cause: High steam velocity in the line.

Consequence: Irritating noise and increased pipe erosion.

Adjustment: Increase the pipe diameter or slow the steam velocity down by installing orifice plates with appropriate sized openings.

Problem: Excessive steam line pressure fluctuations.

Cause: Control valves are opening and closing rapidly or too slowly. This is also many times accompanied by steam line hammer.

Consequence: Excessive pipe movement, nuisance banging noises, and poor steam quality can result from rapid depressurization.

Adjustment: Regulate control valve opening and closing times. If a PRV is the culprit, then a different type valve may have to be used.

Problem: Product is not being heated up fast enough.

Cause: Not enough steam is being supplied to the product or the steam is being supplied at too low a temperature. Steam control valve or steam piping size may be too small. The steam pressure may be too low or the steam-using equipment may be fouled. The condensate may be backing up into the steam-using equipment causing water logging. The product heat exchanger may not have enough surface area.

Consequence: Long cycle times and decreased production.

Adjustment: Increased steam pressure will provide hotter steam. Increasing the pipe size or control valve size will allow more steam to the product. Increasing the size of the steam trap or relocating the steam trap farther from the steam-using equipment will prevent water logging. If the steam-using equipment is fouled, then performance will only be improved by removing the fouling condition. If none of these help, then upsize the heat exchange to increase the surface area.

Problem: Product is being overheated.

Cause: Poor selection or adjustment of the steam control valve.

Consequence: Product ruined.

Adjustment: A control valve replacement with a type that will provide tighter control will help. Lowering the steam temperature (i.e., pressure) to the product will reduce temperature of the product. Verify temperature sensor is correct and functioning properly.

Problem: Loud banging noises from steam piping.

Cause: Water hammer caused by hot condensate slug being pushed at high speed down the steam line until it hits a control valve or pipe changing direction. Hammer could also be caused by rapid collapse of two phase flow in a condensate line.

Consequence: Pipe rupture and nuisance noise.

Adjustment: Add drip leg to remove condensate buildup in the steam lines. Insulate the steam line. Slow down opening or closing of the control valve. In the case of condensate line hammer introduce hot condensate via a diffuser into the main condensate line.

21

COMMISSIONING AND TROUBLESHOOTING THE CONDENSATE AND FEED WATER SYSTEM

When steam surrenders its energy, it will condense and need to be collected, discarded, or pumped back into the boiler. The condensate and feed water system includes condensate collection equipment and feed water equipment. This part of the steam system has three fundamental purposes. It must collect the condensate, treat the condensate to make quality feed water, and provide a means to pump the feed water back to the boiler. A well-designed condensate and feed water system can perform these functions without wasting water and water treatment chemicals despite a range of condensate flows and temperatures. **The Condensate and Feed Water Commissioning Data Sheet** in Appendix B can be used to document system adequacy.

Condensate Collection

The trapped condensate must be collected and stored as feed water for the boiler. A typical condensate collection system will consist of one or more of the following components: flash tank, condensate receiver, pressure motive pump, and surge tank. Condensate collection will vary with the size of the system and may not use all of the components listed above. The sole purpose of the condensate collection system is to reduce the pressure/temperature of the condensate, collect it, and feed it to the feed water system. Commissioning engineers need to ensure the condensate collection system adequately serves all three functions.

Process Steam Systems: A Practical Guide for Operators, Maintainers, Designers, and Educators, Second Edition. Carey Merritt.
© 2023 John Wiley & Sons, Inc. Published 2023 by John Wiley & Sons, Inc.

The commissioning engineer should check to see if the condensate line size, flash tank, receivers, and pump traps all have capacities meeting or exceeding the condensate load. Also, as a steam system is allowed to cool down (upon shut down) the steam will condense in the steam piping and sit there until the system is either drained or re-pressurized. Long condensate lines should have automatic dump valves or manual drain valves installed. Large low-pressure steam systems will often see significant amounts of condensate slug back to the feed water or surge tank during startup after a system shutdown. If the tramp condensate is not drained prior to system restart, then severe water hammer can occur. It is considered a good practice to have operators open all the condensate leg drains and drain any condensate prior to restarting the steam system. This is a good reason to not allow a low-pressure steam system to shut down completely at night.

The condensate collection system may include a surge tank. This tank can be an ASME code or noncode tank that is capable of storing condensate and feeding that condensate to the feed water tank. The surge tank size will depend on the steam system but is generally about 50–75% of the volume of the feed water tank. It should be constructed of stainless steel or a type of corrosion resistant material. Cold treated makeup water is often added to the surge tank if the steam system water inventory becomes too low. The surge tank size, level control system material construction, pump adequacy, and logic ties to the feed water system should be verified.

FIGURE 21.1 Maintaining the feed water deaerator.

Feed Water System

A boiler feed water system, similar to what is shown in Figure 21.1, receives the condensate once it is collected, treats the water, and pumps it back to the boiler. Typically the storage volume of the feed water tank or de-aerator will be 10 min of capacity based on rated output mass flow of the boiler(s). Therefore a 200 hp boiler will have a mass flow of 200 bhp × 0.69 gpm/bhp = 13.8 gpm steam flow. Since one pound of steam equals one pound of water, then the feed water tank for this size boiler should have a capacity of at least 13.8 gpm × 10 min = 138 gallons. This should be the working volume, not the full volume. Some void space is needed in the feed water tank to accommodate surges of water, especially at startup of a steam system.

The feed water system needs to prepare the water for injection into the boiler. This preparation includes preheating, deaerating, and treating the feed water. Chemical injection of any water treatment chemicals, if used, will likely occur in the feed water system. The feed water system last function is to pump the hot pretreated water back to the boiler when the boiler calls for water. Feed water pumps must have adequate capacity and NPSHA at the feed water temperature to facilitate smooth even feed water flow into the boiler. ASME code requires the feed water equipment manufactures to use pumps that are rated for the boilers design pressure, not the operating pressure. Feed water pumps are often oversized, undersized, or the wrong type of pump. Commissioning engineers should verify the tanks, pumps, and line sizes are adequate and installed as designed. In addition strainers, pressure and temperature gauges, and check valves are installed in the right direction and location.

Troubleshooting the Condensate and Feed Water Systems

Problem: Excessive feed water pump cycling.

Cause: Feed water flow into the boiler is excessive relative to the water volume needed to maintain water level.

Consequence: Premature failure of the feed water pump motor.

Adjustment: Throttle down on the feed pump discharge valve, add a pump recirculation line with a relief valve discharging back to the feed tank, or replace the pump with a smaller pump.

Problem: Feed water pump does not fill the boiler fast enough to maintain water level.

Cause: The feed pump is too small, the pump is cavitating, the feed water line is blocked, or the pump is oversized and experiencing run out.

Consequence: Nuisance low water shutdowns.

Adjustment: First ensure a pressure gauge is installed in the feed water line relatively close to the boiler. Next, check the pump suction line and discharge line to the boiler is free of debris. Record the pump discharge pressure and verify the pump capacity is adequate using the pump curve found in the O&M manual. Replace the pump with a larger capacity pump if current pump is too small. If the current pump is too large, then throttle the pump discharge flow enough to bring the pump capacity back into the area of the pump curve that is suitable for the boiler pressure. If the pump sounds like it is cavitating (marbles), then verify you have enough suction head to satisfy the requirements of the pump.

Problem: DA cannot keep 3–5 psig pressure.

Cause: The DA steam PRV is not sized right or not adjusted right or the DA is being over vented.

Consequence: The feed water will not be fully deaerated resulting in the boiler and DA being subjected to accelerated oxygen corrosion.

Adjustment: Verify the venting is not excessive. Adjust the PRV to allow more steam through the valve. Using the valve cut sheet, verify the PRV will pass enough steam to keep the DA at 3–5 psig. Replace the PRV if it is too small.

Problem: DA internal inspection shows reddish-orange deposits and metal color.

Cause: Inadequate deaeration of the water in the DA. This is likely caused by insufficient venting of noncondensable gases or inability of the DA to reach saturation temperature.

Consequence: Premature failure of the DA vessel due to excessive oxygen corrosion. Unprotected boiler.

Adjustment: Increase DA venting and/or steam supply from PRV. Ensure DA water temperature is 223–227 F at 3–5 psig pressure.

Problem: Surge tank or feed water tank overflows during system startup.

Cause: Tramp condensate in the system is pushed back to the feed water system after startup.

Consequence: Nuisance overflow and wasted water.

Adjustment: The tramp condensate should be manually drained from all the steam drip legs prior to startup or automatic dump valves can be installed on the drip legs. The level can be manually held in a low condition until after system startup is complete.

Problem: Feed water (DA) or surge tank are being overpressurized with steam.

Cause: Subcooling of the condensate is not occurring. This is caused by leaking steam traps, an open bypass valve, or condensate piping not allowed to cool enough from over insulation or too short a condensate run.

Consequence: Water is wasted, tank damage may occur, and unwanted steam discharge will occur.

Adjustment: Fix the leaking steam traps, close any bypass valves, uninsulate the condensate piping, or make the condensate pipe run longer to the feed water tank.

Problem: DA water level bounces and pumps seem to cavitate when makeup water is introduced to the DA.

Cause: When relatively cold water is added to the DA, the steam blanket collapses and so does the pressure inside the DA. In extreme cases, the DA may actually go to a negative pressure momentarily. This condition causes pressure spikes in the DA which create water level oscillations and feed water pump loss of NPHSA.

Consequence: DA pump and steam control valve premature failure, or a boiler level control problem.

Adjustment: Either preheat the makeup water or throttle the flow rate to allow the DA steam control valve to keep up with the steam blanket collapse rate.

Problem: Feed water pump sounds like it has marbles in it or is very loud.

Cause: Pump cavitation caused by not enough suction head pressure or pump run out caused by very low discharge pressure.

Consequence: Premature failure of the pump internals.

Adjustment: Cool the inlet water temperature or raise the feed water tank if the suction head pressure is too low. If runout is occurring, then, close down the pump discharge valve to put more back pressure on the pump.

Problem: Feed water pump lost flow or discharge pressure over time.

Cause: Worn impellor likely due to chemical degradation or erosion.

Consequence: Boiler trips off-line on low water.

Adjustment: Rebuild or replace the pump.

Problem: Excessive oxygen levels in the deaerator.

Cause: Inadequate deaerating or improper venting.

Consequence: Excessive corrosion in the DA. Unprotected boiler.

Adjustment: Verify steam pressure of 3–5 psig is being maintained in the DA. Make sure the water storage compartment is not stratified. Verify the noncondensable gases are being vented properly. The water temperature should be within 2 F of the saturation temperature for the DA pressure. Inspect DA internals and verify trays or spray nozzles are clear and intact.

Problem: Loud banging in the condensate lines.

Cause: Water hammer due to steam collapsing creating a pressure wave.

Consequence: Pipe rupture or steam trap destruction.

Adjustment: Add condensate diffuser to regulate condensate flow into main condensate lie. Insulate condensate lines.

22

COMMISSIONING AND TROUBLESHOOTING THE WATER TREATMENT EQUIPMENT

Maintaining good heat transfer efficiency and preserving the integrity of the pressure boundary is the reason we use water treatment chemicals. We have seen in Chapter 11 that there are three detrimental internal conditions that can occur in a steam system: scaling, fouling, and corrosion. In addition, improper water chemistry can cause poor steam quality and nuisance shutdowns. Commissioning engineers need to ensure adequate water treatment measures are employed that will prevent the three adverse conditions and ensure high steam quality. Understanding some basic water chemistry principles is a good place to start.

In summary, there are many different methods used to protect the steam system. Water chemistry reports will provide the commissioning engineer with the data that shows what conditions are being maintained in the steam system. The **Water Chemistry Monitoring Data Sheet adopted by the Harford Steam Co**. shown in Appendix C shows the parameters and frequency of monitoring that should be incorporated in all steam system operations. Most owners will use a reputable water treatment company to help set up a responsible program. It is always a good practice to obtain benefit versus chemical concentration curves for any water treatment chemical added to your system. These curves can be used to ensure water treatment chemicals are not over- or underfed.

Setting Up the Water Treatment Systems

Makeup water is any water that is added to the steam system as a result of lost condensate, wasted steam, or unrecovered blowdown water. Softening, deionization, or reverse osmosis systems are used to treat makeup water and should be installed upstream of the feed water deaerator. Chemical feed systems are set up to add water treatment chemicals

Process Steam Systems: A Practical Guide for Operators, Maintainers, Designers, and Educators, Second Edition. Carey Merritt.
© 2023 John Wiley & Sons, Inc. Published 2023 by John Wiley & Sons, Inc.

to the feed water deaerator, the boiler, or the steam piping. Adequate sample connections and coolers should be used to provide places to monitor water chemistry parameters. The commissioning engineer should verify chemical pump sizing, injection point correctness, and sample valves are installed in key locations.

Troubleshooting Water Treatment System Problems

Problem: Severe water level bouncing in the boiler sight glass.

Cause: High TDS, alkalinity, soap, or oil levels in the boiler.

Consequence: Nuisance shutdowns on low water level.

Adjustment: Increase blowdown. In severe cases, the boiler should be shutdown, drained, and reboiled out and then returned to service.

Problem: Boiler is experiencing high carryover of boiler water with the steam.

Cause: The cause for high carryover can be high exit steam velocity, control valve opening speed, or improper water chemistry. Consult the steam generator and steam delivery sections. If the boiler water TDS or alkalinity levels get too high, the boiler will foam and cause high carryover.

Consequence: Poor steam quality, inefficiency of heat transfer to the product, and extra burden on the feed water pumps.

Adjustment: Lower TDS and alkalinity levels in the boiler via increased blowdown, reduction of water alkalinity builder treatment chemical, or install a dealkalizer in the makeup water treatment system.

Problem: Boiler internal inspection shows a reddish brown color on the water side of the pressure vessel.

Cause: Boiler water contains oxygen causing iron corrosion in the boiler *or* iron corrosion products are being carried into the boiler from the feed water system. The reddish color is from iron corroding in the presence of oxygen.

Consequence: High iron corrosion rates in the boiler or elsewhere in the steam system. Premature failure of the equipment corroding and fouling of the boiler will result if this condition is not mitigated.

Adjustment: Find the equipment that is corroding. Start sampling the condensate feed water and boiler water to see where the iron is coming from. The boiler will concentrate the iron and you should be able to use cycles of concentration in the boiler water as a measurement to estimate the expected iron levels based on the feed water quality. Add water treatment chemical to the part of the system corroding to prevent further corrosion. Reperform internal inspection every 3 months until the boiler internals have a dark brown to black color (good form of iron corrosion called magnetite).

Problem: Boiler water analysis shows high levels of calcium and magnesium despite the fact that you have a water softener installed in the makeup waterline.

Cause: Water softener needs more frequent regeneration or the resin needs to be replaced.

Consequence: Boiler pressure vessel scaling and loss of boiler efficiency.

Adjustment: First test the softener effluent to see if the softener resin needs to be replaced. If the sample shows good softener performance, then initiate a softener sampling scheme that will check softener performance at various times of the cycle…, that is, at 100 gallons, 1000 gallons, after regeneration, etc. Replace resin or reset regeneration cycle to a more appropriate cycle time.

Problem: Boiler water shows low or high pH.

Cause: Intrusion of organics in the feed water, poorly performing pH adjustment chemical treatment.

Consequence: Carbon steel corrosion is minimal at water pH of 9.5–11. PH lower than 8.5 can cause general corrosion of the boiler pressure vessel. Increased boiler water iron levels will result. High pH (>12) can also lead to caustic corrosion.

Adjustment: Add caustic and/or alkalinity to bring the pH up to >9.5. Remove or reduce caustic addition to lower the pH.

Problem: Boiler internal inspection shows excessive sludge buildup.

Cause: Inadequate blowdown causing high boiler water TDS.

Consequence: Reduced heat transfer and loss of efficiency.

Adjustment: Add a dispersant to the boiler water and increase the bottom blowdown frequency or duration to remove sludge.

Problem: Boiler water treatment chemical level cannot be maintained.

Cause: Chemical feed system not working or adjusted too low. High boiler blowdown or excessive boiler water carryover.

Consequence: Boiler and steam system in not being fully protected.

Adjustment: Sample the condensate and verify the water treatment chemicals are not being carried over. Adjust chemical feed water pumps for higher output. Replace pumps if their output capacity is too low. Verify chemical feed pump discharge lines are not plugged.

23

SAMPLE PROBLEM SETS

1. Given a DA tank that is 36″ diameter by 60″ long operating at 5 psig steam pressure. How much heat will it lose? No significant air flow around the tank
   a. With a 80 F ambient room temperature if it is uninsulated.
   b. Same if it had 2 inches of glass wool-type insulation.
   c. What is the heat loss of same uninsulated tank with 10FPS air flow around the tank?

2. You have a 100 hp, 120 psig steam boiler being fed by an atmospheric feed water tank. The steam produced by the boiler is used in an autoclave and the condensate is discarded. At rated output, how much extraction steam would need to be injected into the feed water tank to maintain the feed water at 205 F if the makeup water is 50 F?

3. A 750 ft, 8 inch steam line carries 10 psig steam through a series of buildings. It is uninsulated and you are tasked with determining how much energy does this line lose every day. Assume 24 h/day operation.

4. A 400 hp horizontal firetube boiler manufacture indicated that at 125 psig steam pressure, the boiler will lose about 2% of its rated input as radiant heat loss. Determine the efficiency at high fire (100% input) and at low fire (10% input).

5. A 250 hp steam boiler is blowdown every day to a blow off separator. If the boiler is operated at 100 psig,
   a. How much flash steam is vented off the separator if the blowdown volume is 12 gallons per blowdown sequence?
   b. If the blowdown in 30 s and the blow off separator has a 3 inch vent, what is the flash steam velocity in the vent line?
   c. How much practical energy could you recover from the blowdown water before it reaches the blow off separator?

Process Steam Systems: A Practical Guide for Operators, Maintainers, Designers, and Educators, Second Edition. Carey Merritt.
© 2023 John Wiley & Sons, Inc. Published 2023 by John Wiley & Sons, Inc.

6. A steam boiler's stack emissions is tested and you find the following data. What is the combustion efficiency?

   a. At low fire the exhaust was 380 F with a CO_2 level of 8%.

   b. At high fire the exhaust was 490 F with a CO_2 of 9.8%.

7. What is the difference between combustion efficiency and fuel to steam efficiency?

8. Why does low-pressure steam carry more heat transfer capacity per pound than high-pressure steam?

9. If you want to use steam to heat a jacked kettle filled with a material and you need to heat it to 350 F,

   a. What pressure of steam would you need?

   b. If the material had a heat capacity of 1.2 btu/lb-F, how much steam would you need to heat 1000 lb in 30 min from 70 F. Assume the kettle will not lose any radiant heat to the ambient air.

10. You have to design a heat exchanger to heat 20,000 lb in an open top tank 8′ wide × 24′ long × 6′ deep of 10% sulfuric acid solution. How much 50 psig steam do you need?

    a. To heat from 80 to 180 F in 2 h. Neglect any convection heat losses to the environment.

    b. To maintain 180 F when 4000 lb of 316L stainless steel is dipped in it for 10 min three times per hour and there is a 5FPS air flow over the top of the tank.

11. A 150 hp steam boiler is designed for a maximum allowable working pressure (MAWP) of 150 psig. The boiler has a 3 inch steam nozzle. If you wish to operate this boiler at 20 psig, what are the consequences with regard to the size of the steam nozzle. What can be done to mitigate the consequences.

12. A manufacturing plant wishes to use the plant's 60 psig steam system to heat up water for the restrooms. If the restrooms can use up to 20 gpm of 140 F water or a total of 6000 gallons per day, what is the impact on the plants steam supply. The city water supply average ranges from 45 to 75 F.

13. The 300 hp steam boiler cannot maintain a boiler conductivity level of 2500 uS/cm^2 at rated output by just blowing down the boiler once per day. Consequently, you are considering an automatic surface blowdown system. If your feed water water (after chemical additions) shows a conductivity level of 325 uS/cm^2, how much boiler water must be continuously blown down to maintain the 2500 uS/cm^2 conductivity level in the boiler water. Steam system has 50% condensate returns.

14. You have to design a stack vent system for natural gas-fired 600 hp steam boiler with an input of 25 MMbtu/h. If the boiler operates at 100% input, 20% excess air and is expected to have a stack temperature of 420 F

    a. What will be your approximate flue gas flow in ACFM and SCFM?

    b. What other information do you need to model the stack static pressure?

    c. What is a good stack design static pressure target?

15. You have a plant steam system operating at 130 psig and will be installing a steam using piece of equipment that requires 5300 lb/h steam at 60 psig.

    a. What type of steam will be downstream of the pressure regulator: Wet steam, saturated steam, or superheated steam?

    b. What is the level of latent and sensible heat in the lower pressure steam?

16. When designing an on/off feed water pump for a 100 hp boiler operating at 100 psig but trimmed at 150 psig, what is a good target pump flow rate and pump discharge pressure?

17. What is the feed water oxygen concentration in an open feed water tank at 125 F versus at 200 F.

    a. How can you remove the remaining oxygen in the feed water?

18. You know you have a leaking steam trap. You also know the steam trap has an 1/8" orifice. How much steam are you potentially leaking by per hour if the steam pressure is 50 psig? How much steam could you be loosing per month?

19. Why is steam condensate corrosive to carbon steel?

    a. What are the methods of reducing condensate line corrosion?

20. How does a water softener work? Why do you need salt for some water softeners to regenerate?

21. What is the purpose of a flash tank? When do you need to use one?

22. When you turn the gas-fired boiler burner to the on position, the fan starts but the burner does not light off and burner shuts down. What could be the problem(s)?

23. When 600 lb/h of saturated steam at 125 psig condenses to 170 F liquid condensate at atmospheric pressure, how much energy does it give up?

24. Calculate the amount of combustion air required to combust 2000 ft^3 of natural gas in a burner that will use 20% excess air.

25. A feed water deaerator services 2750 hp, 120 psig steam boilers. The boilers are operating at 75% output and 75% condensate return. If the condensate temperature is 195 F,

    a. Calculate the amount of makeup water required in gpm.

    b. Calculate the DA combined water temperature if the makeup water is 65 F with no preheating.

    c. How much steam is needed to bring the feed water up to 227 F?

    d. What will be the DA pressure at 227 F feed water?

26. Your process requires 2400 gallons of a liquid material with a heat capacity of 0.8 btu/lb and a density of 7.6 lb per gal to be heated to 620 F. You have 200 psig saturated steam available.

    a. How can you use 200 psig steam to heat this material.

    b. Calculate the amount of thermal energy per pound you would need to superheat the steam to achieve 700 F steam at 200 psig.

    c. Calculate the amount of electric energy needed to be added to 200 psig steam to get 700 F steam.

27. Your makeup water to the steam system will use a water softener to remove hardness (calcium and magnesium) from the water. If the water softener has a 150,000 grain capacity, how many gallon of makeup water can be safely put through the softener before it needs regeneration if the makeup water has a hardness of 250 ppm.

28. A 1200 ft steam pipe with a diameter of 8 inches is being considered for a plant modification. If you are going to put 120 psig steam in this line, how much will the pipe expand when heated from 70 F to the saturation temperature?

    a. For an A36 grade carbon steel pipe.

    b. For a 304 stainless steel pipe.

29. What are the causes of water hammer and how do you mitigate the hammer?

30. What are the causes of steam hammer and how do you mitigate it?

31. It is often said that a boiler with a high water volume responds better to rapid steam load swings than a low water volume boiler. Why is this true? What thermodynamic principle backs up this claim?

32. A natural gas-fired boiler has a flue gas NOx concentration of 34 ppm, 8.4% CO_2, and 6% O_2. What is the NOx concentration at 3% O_2?

33. Draw the heat conductance chart showing the relative heat conductance through a boiler burner combustion chamber through refractory brick metal wall into the boiler water.

34. What gauges or sensors with indicators are necessary for a well designed steam system?

35. Why do manufactures request at least 10 pipe diameters of straight pipe be located upstream and downstream of a pressure regulator, sensor, or meter?

A

REFERENCES AND REFERENCE INFORMATION

REFERENCE PAGE

1 "The Energy Fluid" has previously been used to describe steam by the Spirax Sarco Co.

2 *Steam Utilization, Design of Fluid Systems*, Spirax Sarco, 1991.

3 *Steam System Best Practices*, Document No. 23, Swagelok Energy Advisors Inc.

4 *A History of the Growth of the Steam-Engine*, Robert H. Thurston, Cornell University Press, Ithaca, NY, originally published 1878; H. Milford, Oxford University Press, London, 1939.

5 *Data extracted from Wikipedia Commons: History of Boiler Explosion Deaths.*

6 *Wikipedia Commons: Boiler Explosion Images, Darlington Explosion*, 1850.

7 Photo originated from the Keene Public Library and the Historical Society, Cheshire County, NH.

8 *Boiler and Pressure Vessel Code*, BPVC 100, An International Code, ASME, 2015.

9 United States Department of Energy, *Cyclic Steam Stimulation Production Process*, USDOE, Washington, DC. Accessed 2021.

10 *Wikipedia: Heat Recovery Steam Generator, IGCC plant diagram.*

11 *North American Combustion Handbook*, Richard J Reed, 3rd Edition, Volume 1, North American Manufacturing Co., 2001.

12 *Perry's Chemical Engineers Handbook*, 6th Edition, McGraw Hill Book Co., 1984.

13 *Thermodynamics Made Simple for Energy Engineers*, S. Bobby Rauf PE, The Fairmont Press Inc., 2012.

14 *Boiler Plant and Distribution System Optimization Manual*, 2nd Edition, Harry Taplin PE, The Fairmont Press Inc., 1998.

15 *Unit Operations of Chemical Engineering*, 3rd Edition, Warren L. McCabe and Julian C. Smith, McGraw Hill Book Co., 1976.

Process Steam Systems: A Practical Guide for Operators, Maintainers, Designers, and Educators, Second Edition. Carey Merritt.
© 2023 John Wiley & Sons, Inc. Published 2023 by John Wiley & Sons, Inc.

16 *Design of Fluid Systems*, 11th Edition, Spirax Sarco Inc., 1997.

17 *Steam, Its Generation and Uses*, 39th Edition, Babcock and Wilcox Co., 1978.

18 *Watson McDaniel Product Catalog*, Watson McDaniel Inc., 2010.

19 *Boiler Efficiency and Steam Quality: The Challenge of Creating Quality Steam Using Existing Boiler Efficiencies*, Glenn Hahn, National Board Bulletin, 1998.

20 *The Boiler Book*, 1st Edition, Cleaver Brooks, Division of Aqua-Chem Inc., 1993.

21 *Design Considerations for Oil System Sizing*, Version 4. Preferred Utilities Manufacturing Corp., 2004.

22 *Steam Filter Stations for Sterilization*, Bulletin SB-P230-01-US-ISS1, www.SpiraxSarco.com.

23 *How to Trap Superheated Steam Lines*, Solution Source for Steam Air and Hot Water Systems, Armstrong International Inc., 2004.

24 *Steam Plant Operation*, 8th Edition, Everett Woodruff, Herbert. Lammers, and Thomas. Lammers. McGraw Hill Inc., 2005.

25 *Improving Steam System Performance: A Sourcebook for Industry*, US Department of Energy, Advanced Manufacturing Office, http://energy.gov/sites/prod/files/2014/05/f15/steamsourcebook.pdf.

26 *Electric Boiler Consultation with Mr. Jerry Porter*, Product Marketing Manager, Precision Boilers, Morristown, TN, 2015.

27 www.Chemicalengineeringworld,com/Heat-Recovery-Steam-Generator-hrsg.

28 Bulletin LF3BUSS02-00EN, *Boiler Control Solutions: Instruments for Automatic Boiler Controls*, Yokogawa Electric Corporation, 2014.

29 What are the Different Types of Biomass Boilers, *The Renewable Energy*, hub.co.uk.

30 *Burner Technology for Single Digit NO_x Emissions in Boiler Applications*, Timothy Wheeler, *Presentation at the CIBO NOx Control XIV Conference*, 2001.

31 *Minimize NOx Emissions Cost Effectively*, J.D. Adams, S.D. Reed, and D.C. Itse, *Hydrocarbon Processing*, June 2001.

32 *NO_x Reduction*, Michael Barnes, Combustion Booklet, E-Instruments International, LLC, 2014.

33 *Source Test Protocol for Determining Oxygen Corrected Pollutant Concentrations from Combustion Sources with High Stack Oxygen Content Based on Carbon Dioxide Emissions*, South Coast Air Quality Management District, March 2011.

34 *Determination of Sulfur Dioxide Removal Efficiency and Particulate Matter, Sulfur Dioxide, and Nitrogen Oxide Emission Rates*, 10CFR40 part 60 Appendix A-7, Test Method 19, 7/1/2009 Edition.

35 *Combustion Engineering Guide*, 6th Edition, Eclipse Inc., 1986.

36 *Steel Pipe Calculating Thermal Expansion Loops*, www.EngineeringToolbox.com.

37 *Types of Steam Meters*, www.SpiraxSarco.com.

38 *Crosby Pressure Relief Valve Engineering Handbook*, Crosby Valve Inc., Technical Document No. TP-V300, 1997.

39 *Handbook of Industrial Water Treatment*, 8th Edition, Betz Inc. 1980.

40 *Corrosion Control in the Chemical Process Industry*, 2nd Edition, C.P. Dillion, MTI Pub. No 45, 1994.

41 *Single Certification Mark*, Reprinted from ASME 2013 BPVC, Section VIII-Division 2, by permission of the American Society of Mechanical Engineers. All rights reserved.

42   *Five Things You Might Have Missed in NFPA-85*, Robert Frohock PE, Today's Boiler Trend, Technology and Innovations, Spring 2014.

43   *How Not to Design a Steam System*, Mr. Paul Pack, The Fulton Co., Engineering Expo, Syracuse, NY, November 2013.

44   *Nation Fuel Gas Code*, NFPA 54, section Z223.1, 2002.

45   *American Society of Heating, Refrigeration and Air Conditioning Engineers (ASHRAE) Handbook-Fundamentals*, 1993.

46   *Combined Heating, Cooling & Power Handbook*, Neil Petchers, The Fairmount Press Inc., 2003.

47   *Consider Installing High Pressure Boilers with Back Pressure Turbine-Generators*, US DOE, Advanced Manufacturing Office, http://energy.gov/sites/prod/files/2014/05/f16/steam22_backpressure.pdf.

48   *Steam Muffler Design*, Mr Tim Coughlin, Senior Process Engineer, C&S companies, 2014.

49   Autoclave Definition, Parts, Principle, Procedures, Types and Uses, Anupama Sapketa, *Microbe Notes*, November 2021.

50   *Wikipedia Commons: Thermosyphon Reboiler*, Author Milton Beychok.

51   *HVAC Equations, Data, and Rules of Thumb*, Aurther A. Bell, Jr. PE, 2nd Edition, McGraw Hill, 2008.

52   *Boiler Efficiency, Facts You Should Know about Firetube Boilers and Boiler Efficiency*, Cleaver Brooks, Division of Aqua-Chem Inc., 1996.

53   Utility Biomass Use: Turning Over a New Leaf?, Una Nowling PE, *Power*, Vol. 158, No 5, 2014.

54   *General Engineering Data*, Boiler Basics, Cleaver Brooks.com, 2014.

55   *Suggested Maintenance Log Program*, Boiler logs Can Reduce Accidents Technical Bulletin, The National Board of Boiler and Pressure Vessel Inspectors, 1995.

56   ASME Section 1 Boiler Code B31.1 power piping.

SATURATED STEAM TABLE [12]

Gauge Pressure (in Hg or psig)	Absolute Pressure (psia)	Temperature (°F)	Specific Volume Steam (ft³/lb)	Heat Content		
				Sensible (hf) (btu/lb)	Latent (hfg) (btu/lb)	Total (hg) (btu/lb)
25″ vacuum		134	142	102	1017	1119
15″ vacuum		179	51.3	147	990	1137
5″ vacuum		203	31.8	171	976	1147
0 psig	14.7	212	26.8	180	970	1150
2	16.7	219	23.5	187	966	1153
4	18.7	224	21.4	192	962	1154
6	20.7	230	19.4	198	959	1157
8	22.7	233	18.4	201	956	1157
10	24.7	239	16.5	207	953	1160

(Continued)

Gauge Pressure (in Hg or psig)	Absolute Pressure (psia)	Temperature (°F)	Specific Volume Steam (ft³/lb)	Heat Content		
				Sensible (hf) (btu/lb)	Latent (hfg) (btu/lb)	Total (hg) (btu/lb)
12	26.7	244	15.3	212	949	1161
14	28.7	248	14.3	216	947	1163
16	30.7	252	13.4	220	944	1164
20	34.7	259	11.9	227	939	1166
24	38.7	265	10.8	233	934	1167
30	44.7	274	9.5	243	929	1172
36	50.7	282	8.4	251	923	1174
40	54.7	286	7.8	256	920	1176
46	60.7	293	7.1	262	915	1177
50	64.7	298	6.7	267	912	1179
60	74.7	307	5.8	277	906	1183
70	84.7	316	5.2	286	898	1184
80	94.7	324	4.7	294	891	1185
90	104.7	331	4.2	302	886	1188
100	114.7	338	3.9	309	880	1189
120	134.7	350	3.3	322	871	1193
135	149.7	358	3.0	330	864	1194
160	174.7	371	2.6	344	853	1197
180	194.7	380	2.3	353	845	1198
200	214.7	388	2.1	362	837	1199
250	264.7	406	1.8	382	820	1202
300	314.7	421	1,5	398	805	1203
350	364.7	435	1.3	414	790	1204
400	414.7	448	1.1	428	777	1205
500	514.7	470	0.9	453	751	1204
600	614.7	489	0.7	475	728	1203
800	814.7	520	0.6	512	686	1198

Data from Perry's *Chemical Engineers Handbook.*

SUPERHEATED STEAM TABLE [12]

Absolute Pressure (psia)	V= Specific Vol. H = Enthalpy	Saturated Steam					Temperature (F)							
			250	300	350	400	500	600	700	800	900	1000	1200	1400
25	V	16.3	16.6	17.8	19.1	20.3	22.7	25.2	27.6	30	32.4	34.7	39.5	
	H	1161	1166	1190	1215	1239	1286	1335	1383	1433	1483	1535	1640	
40	V	10.5		11	11.8	12.6	14.1	15.7	17.2	18.7	20.2	21.7	24.7	
	H	1170		1187	1212	1237	1285	1334	1383	1432	1483	1534	1639	
60	V	7.2		7.3	7.8	8.4	9.4	10.4	11.4	12.4	13.5	14.5	16.5	
	H	1178		1182	1208	1234	1283	1332	1382	1431	1482	1535	1639	
80	V	5.5			5.8	6.2	7	7.8	8.6	9.3	10.1	10.8	12.3	13.8
	H	1183			1205	1231	1281	1331	1381	1431	1481	1533	1639	1747
100	V	4.4			4.6	4.9	5.6	6.2	6.8	7.4	8.1	8.7	9.9	11.1
	H	1187			1200	1228	1279	1330	1379	1430	1481	1532	1638	1747
120	V	3.7			3.8	4.1	4.6	5.2	5.7	6.2	6.7	7.2	8.2	9.2
	H	1190			1196	1225	1277	1328	1378	1429	1480	1532	1638	1747
140	V	3.2				3.5	4	4.4	4.9	5.3	5.7	6.2	7	7.9
	H	1193				1221	1275	1327	1377	1428	1479	1531	1637	1747
160	V	2.8				3	3.4	3.8	4.2	4.6	5	5.4	6.2	6.9
	H	1195				1218	1273	1325	1376	1427	1479	1531	1637	1746
180	V	2.5				2.6	3	3.4	3.8	4.1	4.4	4.8	5.5	6.1
	H	1197				1214	1271	1324	1375	1426	1478	1530	1637	1746
200	V	2.3				2.4	2.7	3.1	3.4	3.7	4	4.3	4.9	5.5
	H	1198				1211	1269	1322	1374	1425	1477	1530	1636	1746

(Continued)

Absolute Pressure (psia)	V= Specific Vol. H = Enthalpy	Saturated Steam	Temperature (F)											
			250	300	350	400	500	600	700	800	900	1000	1200	1400
240	V	1.9				1.9	2.2	2.5	2.8	3.1	3.3	3.6	4.1	4.6
	H	1201				1203	1265	1319	1372	1424	1476	1528	1635	1745
280	V	1.7					1.9	2.2	2.4	2.6	2.8	3.1	3.5	3.9
	H	1203					1260	1316	1370	1422	1475	1527	1635	1745
320	V	1.4					1.7	1.9	2.1	2.3	2.5	2.7	3.1	3.4
	H	1204					1256	1313	1367	1420	1473	1526	1634	1744
360	V	1.3					1.4	1.7	1.8	2	2.2	2.4	2.7	3.1
	H	1205					1251	1310	1365	1419	1472	1525	1633	1743
400	V	1.2					1.3	1.5	1.7	1.8	2	2.1	2.4	2.8
	H	1205					1246	1307	1363	1417	1470	1524	1632	1743

Data from Perry's *Chemical Engineers Handbook*.

Example Saturated steam at 120 psia superheated to 700 F would need 1378 – 1190 = 188 btu/lb energy added.

SATURATED WATER TABLE

Temp	Sat Press	Spec Volume		Spec Int. Energy		Spec Enthalpy	
		ft³/lbm		btu/lbm		btu/lbm	
F°	psia	Sat Liquid	Sat Vapor	Sat Liquid	Sat Vapor	Sat Liquid	Sat Vapor
T	Psat@T	V_f	V_g	U_f	U_g	h_f	h_g
32.018	0.08866	0.016022	3302	0	1021.2	0.01	1075.4
35	0.09992	0.016021	2948	2.99	1022.2	3	1076.7
40	0.12166	0.01602	2445	8.02	1023.9	8.02	1078.9
45	0.14748	0.016021	2037	13.04	1025.5	13.04	1081.1
50	0.17803	0.016024	1704.2	18.06	1027.2	18.06	1083.3
60	0.2563	0.016035	1206.9	28.08	1030.4	28.08	1087.7
70	0.3632	0.016051	867.7	38.09	1033.7	38.09	1092
80	0.5073	0.016073	632.8	48.08	1037	48.09	1096.4
90	0.6988	0.016099	467.7	58.07	1040.2	58.07	1100.7
100	0.9503	0.01613	350	68.04	1043.5	68.05	1105
110	1.2763	0.016166	265.1	78.02	1046.7	78.02	1109.3
120	1.6945	0.016205	203	87.99	1049.9	88	1113.5
130	2.225	0.016247	157.17	97.97	1053	97.98	1117.8
140	2.892	0.016293	122.88	107.95	1056.2	107.96	1121.9
150	3.722	0.016343	96.99	117.95	1059.3	117.96	1126.1
160	4.745	0.016395	77.23	127.94	1062.3	127.96	1130.1
170	5.996	0.01645	62.02	137.95	1065.4	137.97	1134.2
180	7.515	0.016509	50.2	147.97	1068.3	147.99	1138.2
190	9.343	0.01657	40.95	158	1071.3	158.03	1142.1
200	11.529	0.016634	33.63	168.04	1074.2	168.07	1145.9
210	14.125	0.016702	27.82	178.1	1077	178.14	1149.7
212	14.698	0.016716	26.8	180.11	1077.6	180.16	1150.5
220	17.188	0.016772	23.15	188.17	1079.8	188.22	1153.5
230	20.78	0.016845	19.386	198.26	1082.6	198.32	1157.1
240	24.97	0.016922	16.327	208.36	1085.3	208.44	1160.7
250	29.82	0.017001	13.826	218.49	1087.9	218.59	1164.2
260	35.42	0.017084	11.768	228.64	1090.5	228.76	1167.6
270	41.85	0.01717	10.065	238.82	1093	238.95	1170.9
280	49.18	0.017259	8.65	249.02	1095.4	249.18	1174.1
290	57.53	0.017352	7.467	259.25	1097.7	259.44	1177.2
300	66.98	0.017448	6.472	269.52	1100	269.73	1180.2
310	77.64	0.017548	5.632	279.81	1102.1	280.06	1183
320	89.6	0.017652	4.919	290.14	1104.2	290.43	1185.8
330	103	0.01776	4.312	300.51	1106.2	300.43	1188.4
340	117.93	0.017872	3.792	310.91	1108	311.3	1190.8

Data from HVAC Equations, Data, Rules of Thumb [51].

COMBUSTION EFFICIENCY DATA FOR NATURAL GAS-FIRED BOILERS

			Net Stack Temperature (F)							
% Excess Air	% Oxygen	% CO$_2$	300	340	380	420	460	500	540	580
0	0	11.7	14.5	15.2	15.8	16.5	17.1	17.8	18.4	19.1
2.2	0.5	11.4	14.6	15.3	16	16.6	17.3	18	18.6	19.3
4.4	1	11.1	14.8	15.4	16.1	16.8	17.5	18.2	18.8	19.5
6.8	1.5	10.9	14.9	15.6	16.3	17	17.7	18.4	19.1	19.8
9.3	2	10.6	15	15.7	16.4	17.2	17.9	18.6	19.3	20
12	2.5	10.3	15.2	15.9	16.6	17.4	18.1	18.8	19.6	20.3
14.8	3	10	15.3	16.1	16.8	17.6	18.3	19.1	19.8	20.6
17.7	3.5	9.8	15.5	16.2	17	17.8	18.6	19.3	20.1	20.9
20.8	4	9.5	15.6	16.4	17.2	18	18.8	19.6	20.4	21.2
24.1	4.5	9.2	15.8	16.6	17.4	18.3	19.1	19.9	20.7	21.5
27.6	5	8.9	16	16.8	17.7	18.5	19.4	20.2	21.1	21.9
31.4	5.5	8.6	16.2	17.1	17.9	18.8	19.7	20.5	21.4	22.3
35.4	6	8.4	16.4	17.3	18.2	19.1	20	20.9	21.8	22.7
39.6	6.5	8.1	16.6	17.6	18.5	19.4	20.3	21.3	22.2	23.1
44.2	7	7.8	16.9	17.8	18.8	19.8	20.7	21.7	22.6	23.6
49	7.5	7.5	17.1	18.1	19.1	20.1	21.1	22.1	23.1	24.1
54.3	8	7.2	17.4	18.4	19.5	20.5	21.6	22.6	23.6	24.7
60	8.5	7	17.7	18.8	19.9	20.9	22	23.1	24.2	25.2
66.1	9	6.7	18	19.2	20.3	21.4	22.5	23.6	24.8	25.9
72.8	9.5	6.4	18.4	19.6	20.7	21.9	23.1	24.2	25.4	26.6
80	10	6.1	18.8	20	21.2	22.4	23.7	24.9	26.1	27.3

Data from Boiler Plant and Distribution System Optimization Manual [14].

Example A boiler has a stack temperature of 500 F and the room temperature is 80 F. The stack gas analysis shows a CO$_2$ level of 10%. The net stack temperature is 500 – 80= 420 F. At 10% CO$_2$ the combustion efficiency is 100 – 17.6 = 82.4%.

COMBUSTION EFFICIENCY DATA FOR NUMBER 2 FUEL OIL

Heat Loss %			Net Stack Temperature (F)							
% Excess Air	% Oxygen	% CO$_2$	300	340	380	420	460	500	540	580
0	0	15.7	11.5	12.2	12.9	13.7	14.4	15.1	15.8	16.5
2.3	0.5	15.3	11.6	12.4	13.1	13.8	14.6	15.3	16	16.8
4.6	1	15	11.8	12.5	13.3	14	14.8	15.5	16.3	17
7.1	1.5	14.6	11.9	12.7	13.4	14.2	15	15.7	16.5	17.3
9.8	2	14.2	12.1	12.8	13.6	14.4	15.2	16	16.8	17.6
12.5	2.5	13.8	12.2	13	13.8	14.6	15.4	16.2	17	17.8
15.5	3	13.5	12.4	13.2	14	14.8	15.7	16.5	17.3	18.2
18.5	3.5	13.1	12.5	13.4	14.2	15.1	15.9	16.8	17.6	18.5
21.8	4	12.7	12.7	13.6	14.5	15.3	16.2	17.1	18	18.8
25.3	4.5	12.3	12.9	13.8	14.7	15.6	16.5	17.4	18.3	19.2

Heat Loss %			Net Stack Temperature (F)							
% Excess Air	% Oxygen	% CO$_2$	300	340	380	420	460	500	540	580
29	5	12	13.1	14	15	15.9	16.8	17.7	18.7	19.6
32.9	5.5	11.6	13.3	14.3	15.2	16.2	17.1	18.1	19	20
37	6	11.2	13.5	14.5	15.5	16.5	17.5	18.5	19.5	20.4
41.5	6.5	10.8	13.8	14.8	15.8	16.8	17.9	18.9	19.9	20.9
46.2	7	10.5	14.1	15.1	16.2	17.2	18.3	19.3	20.4	21.4
51.4	7.5	10.1	14.3	15.4	16.5	17.6	18.7	19.8	20.9	22
56.9	8	9.7	14.6	15.8	16.9	18	19.2	20.3	21.4	22.5
62.8	8.5	9.3	15	16.1	17.3	18.5	19.7	20.8	22	23.2
69.2	9	9	15.3	16.5	17.8	19	20.2	21.4	22.6	23.9
76.2	9.5	8.6	15.7	17	18.3	19.5	20.8	22.1	23.3	24.6
83.8	10	8.2	16.1	17.5	18.8	20.1	21.4	22.8	24.1	25.4

Data from Boiler Plant and Distribution System Optimization Manual [14].

Example A #2 oil-fired boiler shows a stack temperature of 540 F and a flue gas O$_2$ concentration of 4%. With a boiler room at 80 F, the net stack temperature is 540 – 80 = 460 F. At 4% O$_2$ the combustion efficiency would be 100 – 16.2 = 83.8%.

COMBUSTION EFFICIENCY DATA FOR WOOD, 40% MOISTURE

Heat Loss %			Net Stack Temperature (F)							
% Excess Air	% Oxygen	% CO$_2$	300	340	380	420	450	600	640	680
0	0	20	17	17.7	18.4	19.1	19.6	22.3	23	23.7
2.4	0.5	19.5	17.1	17.8	18.5	19.3	19.8	22.5	23.2	24
4.9	1	19	17.2	17.9	18.7	19.4	20	22.8	23.5	24.3
7.5	1.5	18.6	17.3	18.1	18.9	19.6	20.2	23	23.8	24.6
10.3	2	18.1	17.5	18.3	19	19.8	20.4	23.3	24.1	24.9
13.2	2.5	17.6	17.6	18.4	19.2	20	20.6	23.6	24.4	25.2
16.3	3	17.1	17.8	18.6	19.4	20.2	20.8	23.9	24.7	25.6
19.5	3.5	16.7	17.9	18.8	19.6	20.5	21.1	24.2	25.1	25.9
23	4	16.2	18.1	19	19.8	20.7	21.4	24.6	25.5	26.3
26.6	4.5	15.7	18.3	19.2	20.1	21	21.6	25	25.8	26.7
30.5	5	15.2	18.5	19.4	20.3	21.2	21.9	25.3	26.3	27.2
34.6	5.5	14.8	18.7	19.6	20.6	21.5	22.2	25.8	26.7	27.6
39	6	14.3	18.9	19.9	20.9	21.8	22.6	26.2	27.2	28.2
43.7	6.5	13.8	19.2	20.2	21.2	22.2	22.9	26.7	27.7	28.7
48.7	7	13.3	19.4	20.5	21.5	22.5	23.3	27.2	28.2	29.3
54.1	7.5	12.9	19.7	20.8	21.8	22.9	23.7	27.7	28.8	29.9
59.9	8	12.4	20	21.1	22.2	23.3	24.2	28.3	29.4	30.6
66.2	8.5	11.9	20.3	21.5	22.6	23.8	24.6	29	30.1	31.3
72.9	9	11.4	20.6	21.8	23	24.2	25.1	29.7	30.9	32.1
80.3	9.5	11	21	22.3	23.5	24.8	25.7	30.4	31.6	32.9
88.3	10	10.5	21.4	22.7	24	25.3	26.3	31.2	32.5	33.8

Data from Boiler Plant and Distribution System Optimization Manual [14].

Example A wood boiler has a stack temperature of 600 F and a flue gas CO_2 concentration of 15.7%. The room temperature is 90 F so the net stack temperature is 510 F. The combustion efficiency is between $100 - 21.6 = 78.4\%$ and $100 - 25 = 75\%$. Interpolating between these numbers shows $78.4\% - 75\%/600 - 450\ F = 0.023\%/1\ F$, and $510 - 450 = 60\ F$. So at 510 F net stack temperature the combustion efficiency is $78.4\% - (60 \times 0.023) = 77\%$

STEEL PIPE PROPERTIES

Pipe Size (in)	OD (in)	Schedule	Wall Thickness	ID (in)	Cross-Sectional Area (ft²)	Circumference (ft²/ft)	Capacity at 6 ft/s (gal/min)	Pressure Drop/100 ft (psi)	Weight (lb/ft)
0.5	0.84	10	0.083	0.674	0.0025	0.22	6.67		0.67
		40	0.109	0.622	0.0021	0.22	5.67	14.4	0.85
		80	0.147	0.546	0.0016	0.22	4.38		1.09
0.75	1.05	10	0.083	0.884	0.0043	0.275	11.4		0.86
		40	0.113	0.824	0.0037	0.275	10	10	1.13
		80	0.154	0.742	0.003	0.275	8.07		1.47
1	1.315	10	0.109	1.097	0.0066	0.344	17.7		1.4
		40	0.133	1.049	0.006	0.344	16.1	7.2	1.68
		80	0.179	0.957	0.005	0.344	13.4		2.17
1.25	1.66	10	0.109	1.442	0.0113	0.435	30.5		1.81
		40	0.14	1.38	0.0104	0.435	27.4	5.2	2.27
		80	0.191	1.278	0.009	0.435	23.9		3
1.5	1.9	10	0.109	1.682	0.0154	0.497	41.6		2.09
		40	0.145	1.61	0.0141	0.497	38	4.3	2.72
		80	0.2	1.5	0.0123	0.497	32.9		3.63
2	2.375	10	0.109	2.157	0.0254	0.622	68.3		2.64
		40	0.154	2.067	0.0233	0.622	62.7	3.2	3.65
		80	0.215	1.939	0.0205	0.622	55.2		5.02
2.5	2.875	10	0.12	2.635	0.0379	0.753	102		3.53
		40	0.203	2.469	0.0332	0.753	89.5	2.5	5.79
		80	0.276	2.323	0.02942	0.753	79.2		7.66
3	3.5	10	0.12	3.26	0.058	0.916	156.1		4.33
		40	0.216	3.068	0.0513	0.916	138	1.9	7.58
		80	0.3	2.9	0.0459	0.916	123.3		10.25
4	4.5	10	0.12	4.26	0.099	1.18	266.4		5.61
		40	0.237	4.026	0.0884	1.18	237.6	1.3	10.79
		80	0.337	3.826	0.0799	1.18	214.8		14.98
5	5.563	10	0.134	5.295	0.1529	1.456	411.6		7.77
		40	0.258	5.047	0.139	1.456	373.8	1.1	14.62
		80	0.375	4.813	0.1263	1.456	346.2		20.78
6	6.625	10	0.134	6.357	0.2204	1.734	593.4		9.29
		40	0.28	6.065	0.2006	1.734	540	0.84	18.97
		80	0.432	5.761	0.181	1.734	486.6		28.57

8	8.625	10	0.148	8.329	0.3784	2.258	1018.8		13.4
		40	0.322	7.981	0.3474	2.258	934.2	0.61	28.55
		80	0.5	7.625	0.3171	2.258	853.8		43.39
10	10.75	10	0.165	10.42	0.5922	2.814	1594.8		18.66
		40	0.365	10.02	0.5475	2.814	1476	0.46	40.48
		80	0.5	9.562	0.4987	2.814	1340.4		64.43
12	12.75	10	0.18	12.39	0.8373	3.338	2254.8		24.17
		40	0.406	11.938	0.7773	3.338	2094	0.37	53.52
		80	0.688	11.374	0.7056	3.338	1900.2		88.63
14	14	10	0.25	13.5	0.994	3.665	2676		36.71
		40	0.438	13.124	0.9397	3.665	2532	0.32	63.44
		80	0.75	12.5	0.8522	3.665	2292		106.13
16	16	10	0.25	15.5	1.3104	4.189	3522		42.05
		40	0.5	15	1.2272	4.189	3300	0.29	82.77
		80	0.844	14.312	1.1766	4.189	3168		107.5

Data from *Perry's Chemical Engineering Handbook* [12].

NATURAL GAS LINE CAPACITY

This table shows gas flow capacities of pipes from 1/2″ to 8″ diameter based upon pressure drop of 5″ of the initial gas pressure for 100 ft lengths of pipe. Conversion factors below the table provide for elbows and for lengths of other than 100 ft.

Gas Line Capacities of Natural Gas with Specific Gravity of 0.65

| **Gas Line Capacities (Cu ft/h Through 100 ft Length)** | | | | | | | | | | | |
| **Diameter of Pipe in Inches** | | | | | | | | | | | |
Initial Gas Pressure	1/2″	3/4″	1″	1 1/4″	1 1/2″	2″	2 1/2″	3″	4″	5″	6″	8″
4″ Water	23	52	104	230	358	724	1180	2150	4510	8210	13,400	27,300
5″ Water	26	58	117	257	400	811	1320	2410	5050	9190	14,900	30,600
6″ Water	28	64	128	282	439	889	1450	2640	5540	10,100	16,400	33,500
7″ Water	31	69	139	305	475	962	1570	2860	5990	10,900	17,700	36,300
8″ Water	33	74	148	327	508	1030	1680	3060	6410	11,700	19,000	38,800
9″ Water	35	79	158	347	540	1090	1780	3250	6810	12,400	20,200	41,200
10″ Water	37	83	166	366	569	1150	1880	3430	7180	13,100	21,300	43,500
11″ Water	38	87	175	385	598	1210	1980	3600	7550	13,700	22,300	45,700
12″ Water	40	91	183	402	625	1280	2060	3760	7890	14,300	23,300	47,700
1/2 lb	43	98	197	433	673	1360	2220	4050	8590	15,400	25,100	51,400
3/4 lb	53	121	234	534	831	1680	2750	5000	10,500	19,100	31,100	63,500
1 lb	62	141	282	622	967	1960	3190	5820	12,200	22,200	36,100	73,800
1 1/4 lb	70	159	319	702	1090	2210	3600	6570	13,800	25,000	40,800	83,300
1 1/2 lb	77	175	351	773	1200	2440	3970	7240	15,200	27,600	44,900	91,800
1 3/4 lb	84	191	382	842	1310	2650	4320	7870	16,500	30,000	48,900	99,900

(Continued)

| | **Diameter of Pipe in Inches** | | | | | | | | | | | |
Initial Gas Pressure	1/2″	3/4″	1″	1 1/4″	1 1/2″	2″	2 1/2″	3″	4″	5″	6″	8″
2 lb	91	205	412	906	1410	2850	4660	8480	17,800	32,300	52,600	108,000
2 1/2 lb	103	233	467	1030	1600	3420	5280	9620	20,200	36,700	59,000	122,000
3 lb	114	259	519	1140	1780	3600	5870	10,700	22,400	40,800	66,300	136,000
3 1/2 lb	125	283	568	1250	1940	3940	6420	11,700	24,500	44,600	72,600	148,000
4 lb	135	307	615	1350	2110	4270	6960	12,700	26,600	48,300	78,700	161,000
4 1/2 lb	146	330	661	1460	2260	4580	7480	13,600	28,500	51,900	84,500	173,000
5 lb	155	352	706	1550	2420	4890	7980	14,500	30,500	55,400	90,200	184,000
6 lb	174	395	792	1740	2710	5490	8960	16,300	34,200	62,200	101,000	207,000
8 lb	211	477	957	2110	3280	6640	10,800	19,700	41,300	75,200	122,000	250,000
10 lb	246	556	1120	2460	3820	7730	12,600	23,000	48,200	87,600	143,000	292,000

For each elbow, add the following length in feet to the total length of pipe:

ft/elbow	1	1.3	1.7	2.6	3.3	5.1	6.7	9.3	14	19	24	32

For total lengths other than 100 ft, multiply the figure shown in the table by the factor corresponding to the desired length as follows:

Length of pipe in feet	10	15	25	50	100	150	200	250	300	350	400	500
Multiplier	3.16	2.58	2.00	1.41	1.00	0.817	0.707	0.632	0.577	0.535	0.50	0.447

Data from Cleaver-Brooks Inc. [54].

OIL LINE CAPACITY TABLE

| **#2 Oil Piping Pressure Drop (Viscosity = 40 SSU and Specific Gravity = 0.9)** | | | | | | | | |
| Fuel Oil Flow Rate (GPH) | **Nominal Pipe Size (Inches)** | | | | | | | |
	0.5	0.75	1	1.5	2	2.5	3	4
25	0.3	0.1	0.0*	0.01	0.00*	0.001	0.000*	0.000*
50	0.6	0.2	0.1	0.01	0	0.002	0.001	0.000*
75	0.9	0.3	0.1	0.02	0.01	0.003	0.001	0.000*
100	1.1	0.4	0.1	0.03	0.01	0.005	0.002	0.001
150	3.6	0.9	0.2	0.04	0.01	0.007	0.003	0.001
200	6	1.6	0.5	0.05	0.02	0.009	0.004	0.001
250	8.9	2.3	0.7	0.06	0.02	0.011	0.005	0.002
300	12.3	3.2	1	0.13	0.03	0.014	0.006	0.002
400	20.3	5.3	1.7	0.22	0.007	0.018	0.008	0.003
500	30.5	7.9	2.5	0.32	0.1	0.042	0.01	0.003
600	42.2	11	3.5	0.45	0.13	0.058	0.02	0.004
700	55.6	14.5	4.6	0.59	0.18	0.076	0.027	0.007

Data from Cleaver-Brooks Inc. [54].

* indicates level is less than 0.001 psi/100ft of line.

GUIDE FOR CAPACITY OF STEAM MAINS

		Capacity of Sch. 80 Pipe in lb/h Steam													
Pressure (psi)	Velocity (ft/s)	1/2"	3/4"	1"	1 1/4"	1 1/2"	2"	2 1/2"	3"	4"	5"	6"	8"	10"	12"
5	50	12	26	45	70	100	190	280	410	760	1250	1770	3100	5000	7100
	80	19	45	75	115	170	300	490	710	1250	1800	2700	5200	7600	11000
	120	29	60	110	175	245	460	700	1000	1800	2900	4000	7500	12,000	16500
10	50	15	35	55	88	130	240	365	550	950	1500	2200	3770	6160	8500
	80	24	52	95	150	210	380	600	900	1500	2400	3300	5900	9700	13000
	120	35	72	135	210	330	590	850	1250	2200	3400	4800	9000	14,400	20500
20	50	21	47	92	123	185	320	520	740	1340	1980	2900	5300	8000	11500
	80	32	70	120	190	260	520	810	1100	1900	3100	4500	8400	13,200	18300
	120	50	105	190	300	440	840	1250	1720	3100	4850	6750	13,000	19,800	28000
30	50	26	56	100	160	230	420	650	950	1650	2600	3650	6500	10,500	14500
	80	42	94	155	270	360	655	950	1460	2700	3900	5600	10,700	16,500	23500
	120	62	130	240	370	570	990	1550	2100	3950	6100	8700	16,000	25,000	35000
40	50	32	75	120	190	260	505	790	1100	1900	3100	4200	8200	12,800	18000
	80	51	110	195	300	445	840	1250	1800	3120	4900	6800	13,400	20,300	28300
	120	75	160	290	460	660	1100	1900	2700	4700	7500	111,000	19,400	30,500	42500
60	50	43	95	160	250	360	650	1000	1470	2700	3900	5700	10,700	16,500	24000
	80	65	140	250	400	600	1000	1650	2400	4400	6500	9500	17,500	27,200	38500
	120	102	240	410	610	950	1660	2600	3800	6500	10,300	14,700	26,400	41,000	58000
80	50	53	120	215	315	460	870	1300	1900	3200	5200	7000	13,700	21,200	29500
	80	85	190	320	500	730	1300	2100	3000	5000	8400	12,200	21,000	33,800	47500
	120	130	290	500	750	1100	1900	3000	4200	7800	12,000	17,500	30,600	51,600	71700
100	50	63	130	240	360	570	980	1550	2100	4000	6100	8800	16,300	26,500	35500
	80	102	240	400	610	950	1660	2550	3700	6500	10,200	14,600	26,000	41,000	57300

(Continued)

Capacity of Sch. 80 Pipe in lb/h Steam

Pressure (psi)	Velocity (ft/s)	1/2"	3/4"	1"	1 1/4"	1 1/2"	2"	2 1/2"	3"	4"	5"	6"	8"	10"	12"
	120	150	350	600	900	1370	2400	3700	5000	9100	15,000	21,600	38,000	61,500	86300
	50	74	160	290	440	660	1100	1850	2600	4600	7000	10,500	18,600	29,200	41000
120	80	120	270	450	710	1303	1800	2800	4150	7200	11,600	16,500	29,200	48,000	73800
	120	175	400	680	1060	1520	2850	4300	6500	10,700	17,500	26,000	44,300	70,200	12000
	50	90	208	340	550	820	1380	2230	3220	5500	8800	12,900	22,000	35,600	50000
150	80	745	320	570	900	1250	2200	3400	4900	8500	14,000	20,000	35,500	57,500	49800
	120	215	450	850	1280	1890	3400	5300	7500	13,400	20,600	30,000	55,500	85,500	120000
200	50	110	265	450	680	1020	1780	2800	4120	7100	11,500	16,300	28,500	45,300	64000
	80	180	510	700	1100	1560	2910	4400	6600	11,000	18,000	26,600	46,000	72,300	100000
	120	250	600	1100	1630	2400	4350	6800	9400	16,900	25,900	37,000	70,600	10,900	152000

Data from Spirax Sarco [16].

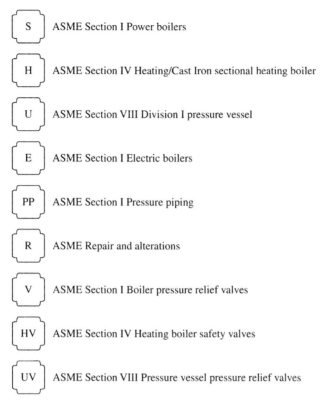

FIGURE A.1   ASME Stamps/Certifications. Data taken from ASME Boiler & Pressure Vessel Code [8].

B

OPERATIONS, MAINTENANCE, AND INSPECTION GUIDANCE

STEAM BOILER DAILY LOG SHEET

Date	Boiler Status On or Off	Boiler Pressure psi	Flame Signal 0 to 5	Firing Rate %	TDS Level ppm	Stack Temperature F	Low Water Cutoff Checks Y/N	Comments

Process Steam Systems: A Practical Guide for Operators, Maintainers, Designers, and Educators,
Second Edition. Carey Merritt.
© 2023 John Wiley & Sons, Inc. Published 2023 by John Wiley & Sons, Inc.

SUGGESTED MAINTENANCE LOG PROGRAM [55]

The following maintenance items, as appropriate to the specific boiler system, need to be considered for implementation on a regular basis (e.g., daily, weekly, monthly, semiannually, annually). A checklist of the items should be incorporated into a maintenance log with provisions for checking off the item for the appropriate period. A separate log sheet is suggested for each period. The log sheets can be filed in a loose-leaf binder and should be retained as a permanent maintenance record.

The log sheets can be used as a handy check-off system when establishing a facility maintenance program.

In all cases the equipment manufacturer's recommendations should be followed.

Daily

- Blowdown and test low water cutoffs of steam boilers (once per shift for high pressure)
- Blowdown gauge glasses (steam)
- Blowdown makeup feeder (low-pressure steam)
- Blow down boiler (steam)
- Check boiler control linkage
- Check boiler and system for leaks
- Check burner flame

Weekly

- Check compressor(s) lubricating oil level (control and atomizing)
- Check flame signal strength for both pilot and main flame, and record readings
- Check flame failure cutoff and timing
- Check pilot and main flame fuel shutoff valve closing
- Check igniter and burner operation
- Check level in chemical treatment tank

Monthly

- Check compressor(s) air filter, and clean or replace as required
- Check boiler water treatment test results received from treatment company, adjust treatment as required
- Lubricate motor and equipment bearings
- Test fan and air pressure interlocks
- Check main burner fuel safety shutoff valves for leakage

- Check low-fire start interlock
- Check high-pressure/temperature interlocks
- Test low water cutoffs (hot water)
- For oil – test pressure and temperature interlocks
- For gas – test high and low gas pressure interlocks
- Manually lift safety/safety relief valves and check operation

Semiannually

- Inspect burner components
- Check flame failure system components
- Check piping and wiring of all interlocks and shutoff valves
- Recalibrate all indicating and recording gauges and instruments
- Perform a slow drain test for low water cutoffs (steam)
- Check combustion control system
- For oil – check atomizers and strainers
- Test boiler safety/safety relief valves in accordance with *ASME Boiler and Pressure Vessel Code, Sections VI and VII*

Annually

- Perform the *semiannual* maintenance procedures
- Check all equipment coils and diaphragms
- Perform a pilot turndown test
- Recondition or replace low water cutoff
- For gas – check drip leg and gas strainer
- Clean boiler firesides
- Drain boiler, open manholes and handholes, and clean watersides
- Have boiler inspected by a commissioned inspector
- Clean burner and sight glass
- Replace gaskets
- Leak-test all fuel valves
- Test operation of all controls and safety devices
- Have fuel-burning system adjusted using combustion test instruments

After Each Period

- Make a record of all maintenance and parts replacement in the maintenance log

SAFETY RELIEF VALVE INSPECTION/VERIFICATION TEST

A. Valve Removal and Inspection

Relief valve _____ Set pressure _____psig

Equipment serviced _____ Date removed _____

Inlet piping condition inspection _____

Discharge piping condition inspection _____

_____

Inspector_____ Date_____

B. Valve Testing

Testing lab ID_____

Testing date _____

"As Received" pretest pop test result _____psig

Record any adjustments or maintenance performed on valve _____

_____

_____

Retest pop test results _____psig

Tester_____Date_____

Staff comments (i.e., valve status, PM frequency change, etc.) _____

_____

BOIL-OUT PROCEDURE

We recommend boil-out be accomplished prior to boiler system operation. This procedure ensures that all oils, sealants, and other organic compounds are solubilized and removed from the boiler and piping. Consequently, if boil-out is not accomplished prior to system operation, erratic water level control and carryover may occur.

Boil-Out Procedure: (For Boiler Only)

1. Fill boiler/return system with water and energize boiler/burner until boiler starts to build steam pressure.
2. Turn off boiler power and lock out control panel in "off position."
3. Open blowdown valve and drain boiler completely. Ensure feed water pump is shut off.
4. Mix boil-out chemical (soda ash or OxiClean) at a rate of 0.5 oz per BHP in a container with water.
5. Remove safety valve, surface blowdown connection, steam stop *or* feed water line opening and pour, *or* pump into boiler. Reinstall boiler system integrity.
6. Reenergize boiler control panel and feed water pump to allow boiler to fill to normal water level.
7. Turn on boiler and allow to come up to at least 10 psig.

8. Turn off boiler and allow steam pressure to dissipate to <5 psig. Allow boiler to sit idle for at least ten minutes.

9. Open blowdown valve and drain boiler. Turn feed water pump to "off position."

10. Refill boiler and repeat steps 6–9.

Boil-Out Procedure (for Entire Steam System)

11. Ensure steps 1–10 have been completed.

12. Add 8–16 oz of boil-out chemical to the DA or return system tank. Ensure chemical is fully dissolved prior to adding to the tank.

13. Turn on boiler and operate under normal operating conditions.

14. Valve in steam loads one at a time to ensure each load has good steam flow. Bypassing steam traps can accelerate system boil-out. However, be careful not to blow raw steam back into the return tank.

15. Once all loads have been valved in and operated for a few minutes, shut the boiler off and allow steam pressure to lower to zero.

16. De-energize control panels and drain entire system. Refill and repeat with 12–15 once more boil-out chemical.

17. Perform a final drain on the system, refill, and place system in operation.

BOILER LAYUP IN STANDBY CONDITION

This guidance has been prepared to summarize the methods that may be followed to prevent or minimize deterioration of the internal surfaces of boilers from corrosion during inoperative periods.

Two sets of conditions require layup:

1. The boiler must be held in readiness to operate at any time on short notice. This may be designated as *intermittent standby and requires a wet layup*.

2. The boiler will be idle for an indefinite period of weeks or months. This is *prolonged standby and requires a dry layup*.

Intermittent Standby

If the feed water is sufficiently oxygen free and of suitable alkalinity (pH value), conditions leading to corrosion will not be developed. However, if the conditions are anything but ideal, low alkalinity or oxygen-rich water may form at the boiler surfaces and initiate corrosion. No single rule can be given to assure correct conditions in the boiler. The regular boiler water tests must be made as carefully on these boilers as on the operating boilers. This is due to the fact that these boilers while in standby cannot be adjusted as easily as the operating boilers, and as such, they can be severely damaged by water problems more easily.

If the alkalinity falls to low, it can be boosted by putting a small amount of alkali solution (preferably sodium bicarbonate) directly into the boiler at the point of feed water entry with a pump or any other convenient manner.

In the cases where quantities of oxygen are dissolved in the feed water, a solution of sodium sulfite can be fed to the boiler by means of a pump or either separately or in conjunction with the alkali solution. The minimum amount of residual sodium sulfite that should be maintained in the boiler is 30–50 ppm as a residual.

The rules for treating a standby boiler are in large the same as those for an operating boiler. The application of chemical treatment to a standby boiler must be as fully and more carefully performed.

Boiler Filled with Treated Water   In this method, protection can be obtained for the boiler metal if

1. Correct chemical conditions are maintained in the boiler water.
2. The boiler water is mixed adequately to maintain uniform conditions throughout the boiler.
3. The boilers are completely filled with **treated water** as not to allow any boiler metal surfaces to come in contact with the air.

Sufficient caustic soda (or equivalent alkalinity builder) should be added to the boiler water in order to produce a hydroxyl (OH–) alkalinity of *300–400 ppm*. In addition, sufficient sodium sulfite must be added to the boiler water to establish a *sodium sulfite residual of 30–50 ppm* as a minimum.

The mixing of the water in the boiler must be thorough so that correct chemical conditions can exist in every section of the boiler. Mixing of the water can be accomplished by circulating water from one section of the boiler to another with the use of a pump, as in prolonged standby, the boiler is not fired off, and as such, cannot mix the boiler water by means of circulation through heating up the boiler. If the pump method is practical, additional chemicals can be trickled into the boiler as they are needed during the circulation procedure.

Deaerated water should be used to fill the boilers that are going to be in standby service, when even it is available. A convenient method for keeping a boiler full of water is to connect a small expansion tank to some connection at the top of the boiler. This tank is to be located above the boiler and is to be kept filled with chemically treated water. If the tank remains full, then the boiler will remain full. If the tank overflows, then the boiler is taking on water from some form of system leakage. It is convenient for maintaining the proper level of chemical treatment protection as well as acting as a level control indicator.

Prolonged Standby

Draining and Drying   This method will allow excellent protection from corrosion to the metal surfaces so long as there is no moisture present in the boiler. One method is to let the boiler open for free circulation of air after drying. Another method is to place a desiccant as silica gel in the boiler and the boiler is then closed up for drying. In either case, water leakage over, or sweating of, the boiler metal surfaces must be protected against. Since this type of moisture is saturated with oxygen, its contact with metal surfaces will cause accelerated rates of corrosion. So long as the boiler metal surfaces remain free of moisture, no appreciable corrosion will occur.

If the boiler's past history indicates that external sweating of the boiler internals is a problem, then additional heater (i.e., light bulb) should be installed in the furnace at strategic locations in order to insure that the boiler pressure vessel is maintained above the dew point. This concept must also include the superheater section if the boiler has a superheater. If you allow dew to form, then the purpose of layup is defeated, as the boiler tube metal will rapidly corrode both internally as well as externally.

For extended dry standby, blanketing with nitrogen should be considered. This method will dry out the metal and if maintained properly, it will help to maintain a moisture-free atmosphere on the internal areas of the boiler.

BOILER SOAKING PROCEDURE

Cold Start

When you start the boiler it is important to minimize the thermal and mechanical stresses applied to the pressure vessel due to the expansion differential of the tubes, furnace, shell, and tube sheets as they reach-operating temperatures. Heating the boiler from cold to quickly during cold startup will diminish the life expectancy of the boiler.

When starting from cold the furnace reaches operating temperature rapidly where the later passes and shell are slower to heat due to the time it takes to heat the boiler water. As a result the thermal expansion of the furnace is the greatest. The temperature gradient between the top of the boiler and the bottom is substantial.

In practical terms the ideal solution would be to gradually raise the temperature and pressure in the boiler progressively. Firing the boiler on low fire for a 10–15 min and then shutting the burner off and allowing the water to "soak" for a period of 10–15 min helps accomplish this. This allows the boiler steel and water temperatures to even out from diffusion during the soak periods. Verification of the heat-up can be obtained by using an infrared heat gun and checking the shell temperature at a handhole or manway. This alternating low firing followed by soaking periods should continue until the shell reaches over 200 degrees or you start to build steam pressure. The heat-up period may take a few hours depending on boiler size. A good rule of thumb is to use 30 min warm-up period per 100 BHP. A 800 hp boiler would require at least 4 h of warm-up time.

If the boiler has been fired the previous day the large thermal mass of the boiler will likely maintain the internal temperature to a point that the boiler will require little warm-up. Generally, it can be switched on and left to reach operating pressure.

The life expectancy of the boiler and door insulation, ignoring other factors, is proportional to the number of thermal, mechanical cycles that the boiler undergoes from cold to working pressure. A boiler that is continually maintained at operating pressure will last longer than a boiler that is constantly heated and cooled. This can be accomplished with the use of optional nighttime setback feature available as a factory-installed option. With a redundant operating pressure control set at lower pressure allows the boiler to operate at low pressure, for example 10 psi. Operating at low pressure and switching to high pressure in the morning is facilitated by a selector switch. This option greatly reduces the time it takes to get to full pressure and eliminates thermal and mechanical stresses on the boiler.

BOILER PERFORMANCE TEST PROCEDURE USING WATER USAGE METHOD

Prerequisites

1. Designate a test director. Verify all ordered equipment is installed and ready to be tested.
2. Verify means of throttling the steam outlet.
3. Connect all utilities.
4. Electrically power up main control panel and subpanels.
5. Place boiler feed pump switch to the "off position."
6. Install boiler stacking if possible.
7. Install all sight glasses and open water column valves (close drain valves).
8. Set pressure control/temperature control high limit and operating limits to meet customer field conditions, if known; otherwise set to 90% of safety valve set point.
9. Plug any open drain connections or close drain/blowdown valves.
10. Shut steam supply valve to the feed water tank or DA and steam stop valve.
11. Close off blowdown separator cooling valve.
12. Place water treatment equipment in bypass.
13. Verify a means to measure:
    a. Fuel input
    b. Incoming fuel pressure
    c. Last elbow gas pressure
    d. Stack gas analysis
    e. Steam output (inlet water consumption method can be used)
    f. Feed water pump discharge pressure
    g. Boiler steam pressure
    h. Incoming water pressure
    i. Incoming water flow

14. Open feed water line valves (duplex systems will line up one pump line at a time).

15. Open fuel supply line.

16. Open boiler vent at steam gauge assembly to allow air to be displaced during boiler filling.

17. Open the return tank/deaerator water supply valve with the panel box powered up and allow the return tank/deaerator to fill up. Record incoming water line pressure.

Boiler Skid Performance Test

1. Close steam stop valve. Turn boiler "on" and warm-up boiler until a couple psig steam pressure is achieved.

2. With the boiler set at low fire adjust on the steam outlet to keep boiler operating within 10 psig of anticipated operating pressure. Allow the boiler to dump steam and operate this way until you are confident that it can be maintained for an hour.

3. With boiler firing at low fire output, determine steam output by measuring the water input (water meter) over a 30 min or more time frame. No blowdown or water leakage can occur during this part of the test. It is best to start to test cycle after the feed water tank is at its highest of lowest water level mark, and conclude the test at the same water level indication. The more water level makeup cycles you obtain (i.e., test duration), the more accurate the test will be. For on/off make up systems, you should operate through at least 10 makeup water cycles in the feed water tank.

4. Record fuel input usage/unit time and fuel pressures at the last elbow or as appropriate if not previously recorded at test fire.

5. Repeat steps 2, 3, and 4 for the high fire output.

6. Using the water usage data from the water meter, you can determine the lowest and highest steam output using the formula

lb water used/min × 60 min/h = lb of steam produced/h and the actual boiler output in btu/h = lb/h steam × (steam temp – feed water temp × 1 btu/lb) + latent heat (btu at the steam temp)

The actual output in btu/h should be within 90% of the advertised rating on the boiler (for high fire).

C

STEAM SYSTEM DESIGN AND COMMISSIONING GUIDANCE

PRECOMMISSIONING CHECKLIST

The following activities should be completed prior to attempting to commission the steam system.

1. **Boiler installation**
   - _____ Ensure SRV and SRV tail pipes on each boiler vented to outside boiler room.
   - _____ Ensure sight glass assemblies and pressure gauges are installed on each boiler.
   - _____ Ensure combustion air ducting and hot exhaust venting are installed.
   - _____ Ensure fuel train assemblies are installed on each boiler.
   - _____ Ensure Blowoff valves/piping is installed on each boiler to blowoff separator.
   - _____ Ensure main power supply to boiler panel(s) is installed.
   - _____ Ensure steam outlet valves and piping are installed to main steam header.
   - _____ Ensure boiler feed water line is piped to the boiler(s).

2. **Feed water tank installation**
   - _____ Ensure tank sight glass and thermometer assemblies are installed.
   - _____ Ensure level control system is installed.
   - _____ Ensure preheat steam source/controls are installed.
   - _____ Ensure condensate return piping to the tank is installed to tank.

Process Steam Systems: A Practical Guide for Operators, Maintainers, Designers, and Educators, Second Edition. Carey Merritt.
© 2023 John Wiley & Sons, Inc. Published 2023 by John Wiley & Sons, Inc.

_____ Ensure tank is vented to atmosphere (for DA make sure you can see vent plume).

_____ Ensure feed water pumps and controls are installed and piped to the boiler.

_____ Ensure make up water source is piped to the tank.

3. **Blowoff separator installation**

_____ Ensure blowoff separator vent line is installed in a safe manner.

_____ Ensure blowoff separator drain line is installed to a safe location.

_____ Ensure blowoff drain water is cooled to <140 F, manually or automatically.

4. **Water treatment system installation**

_____ Ensure boiler blowdown scheme will keep the boiler TDS at or below the warranty limits.

_____ Ensure the boiler has ample corrosion, fouling, and scaling protection via chemical or nonchemical means.

_____ Ensure there is a boiler sample valve installed.

5. **Steam delivery system installation**

_____ Ensure steam boiler outlets are piped to the main steam header.

_____ Ensure all valves and sensors are installed in the steam piping.

_____ Ensure there is a means to dump steam for boiler testing and over pressure protection.

6. **Condensate system installation**

_____ Ensure all steam-using equipment is trapped.

_____ Ensure the condensate return lines all cascade to a collection tank or the condensate is discarded.

_____ Ensure there are sample location installed in the condensate lines.

STEAM SYSTEM COMMISSIONING DATA SHEETS

Date_____ Job No._____ Owner/Client_____

Commissioning Engineer_____

1. **As-found conditions:** This is where the material condition, room ambient conditions, installation discrepancies, workmanship, safety concerns, utility service concerns, piping orientation concerns, missing equipment, and any other system performance concerns are documented.

_____

_____

_____

_____

_____
_____
_____
_____
_____

2. **Commissioning the steam generator**
   a. Verify proper fuel train/supply for rated load.
   b. Verify combustion creates blue (gas) yellow- blue flame (oil) for high and low fire outputs.
   c. Verify pre- and postpurge times are per manufacturer's recommendations.
   d. Use test fire data or a perform a performance test to verify rated output is within 10% of nameplate stamping.
   e. Verify SRV settings are less than or equal to generator MAWP.
   f. Verify water level remains visible in the sight glass during operation.
   g. Verify low water cutoff, high-pressure, combustion air, and fuel pressure switches have been tested satisfactory.
   h. Verify the steam outlet will yield steam flows <5000 ft/min at rated output.
   i. Verify blowdown scheme will keep the boiler solids level within warranty limits.
   j. Verify each generator connected to a common steam header has high-water protection.
   k. Verify the operating pressure switch is set at least 10% below the high-pressure switch set point.
   l. Verify the steam generator will be able to build and maintain the desired pressure during normal and peak load operation.

3. **Commissioning the steam delivery system**
   a. Verify the steam main is sized for steam velocities 6000–8000 ft/min.
   b. Verify the steam main has a trapped drip leg installed.
   c. Verify the steam main has a vacuum breaker/air vent installed.
   d. Verify PRV stations have been installed with all the recommended equipment.
   e. Verify any installed sensors are located at least 10 pipe diameters from a tee, elbow, or valve.
   f. Verify the steam mains are sloped in the direction of steam flow.
   g. Verify each piece of steam-using equipment has an isolation valve.

4. **Commissioning the feed water system**
   a. Verify the tank has at least 10 min worth of water storage based on rated boiler(s) output.

   b. Verify the feed water tank (for pressurized DA only) is being maintained at 3–5 psig and the stored water temperature corresponds to that pressure.

   c. Record the feed water tank temperature (for atmospheric tanks) and determine the residual oxygen concentration.

   d. Verify the residual oxygen is rendered harmless by water treatment chemical additions.

   e. Verify the makeup water level control system keeps the feed water tank level within the sight glass visible range.

   f. Verify the feed water pump(s) operate without cavitation and are able to keep the boiler(s) full of water.

   g. Verify the feed water pump discharge pressure puts the pump operation in the correct part of the pump curve.

   h. If the feed water tank is equipped with a steam injection system, verify the steam injection is adjusted to keep the tank >190 F and <210 F.

   i. Verify the tank can be drained to a safe location.

5. **Commissioning the condensate system**

   a. Verify condensate traps of the right size and type are installed for each piece of steam-using equipment.

   b. Verify there are condensate drain valves installed that will allow cold condensate to be drained during startup periods.

   c. Verify the steam traps are functioning properly and not causing water logging of the steam equipment.

   d. Verify the condensate receivers and pumps are functioning properly.

   e. Verify the surge tank (if installed) is logic tied to the feed water tank and level control is being maintained in both.

   f. Verify adequate temperature sensors/gauges are installed to allow for steam trap performance monitoring.

6. **Commissioning the water treatment systems**

   a. Verify the boiler blowdown scheme is keeping the boiler solids level within warranty limits.

   b. Verify the boiler oxygen level is being maintained at <0.005 cc/l. This can be done by showing a residual level of oxygen scavenging chemical in the boiler.

   c. Verify the chemical feed pumps and tanks are installed and being operated by competent operators.

   d. Verify the chemical feed pumps are delivering enough water treatment chemicals to the steam system to protect the equipment.

   e. Verify each chemical storage tank is clearly labeled with the chemical contained in the tank.

   f. Verify there is a safety shower/eyewash station in the vicinity of the chemical treatment area.

STEAM SYSTEM DESIGN CHECKLIST

Steam Load

1. Calculated load btus/34,500 btu/bhp = _____ bhp
2. Pick up losses (30%), total steam requirement is bhp × 1.3 = _____ bhp

Boiler Design Specifications

1. Pressure vessel MAWP _____psig Operating pressure_____ psig
2. Heating surface area _____ft^2/bhp (>3 ft^2/bhp recommneded0
3. Water content _____gallons at normal full level
4. Steam exit velocity _____ (<5000 ft/min at the operating pressure)
5. Boiler type _____
6. Number of boilers _____ (if >1, recommend a lead lag system)
7. Boiler external piping hydro test required? _____ Yes _____ No
8. SRV's capacity/set points _____ @ _____ pressure

Burner/Input Requirements

1. Fuel type _____(NG/LP/#2 oil/#6 oil/coal/wood chips, electrical, other)
2. Fuel usage at rated output _____ @ pressure _____
3. Boiler turndown _____
4. Combustion air requirement _____ CFH
5. Fuel/air control _____ (i.e., mechanical linkage, servomotors, or fixed)
6. Boiler vent pressure requirement _____ inches WC at rated output

Boiler Room Design Considerations

1. Electrical power required _____ amps @ _____ voltage
2. Fresh air openings _____ ft^2 (1″/4000 btu input)_____N/A, sealed combustion
3. Fuel line venting_____ # lines @ _____ size
4. Water supply pressure and flow rate requirement _____ gpm @ _____ psig
5. Fuel line size required _____
6. Boiler vent classification (i.e., I,II,III,IV)_____, size _____ in
7. Vent construction _____ Rise/run _____

8. Boiler blowdown rate
   a. Condensate return % _____
   b. Boiler estimated blowdown rate _____ gpm
   c. Cooling water requirement _____gpm
9. Boiler room heat generation calculation
   a. Boiler radiant heat losses _____btu/h
   b. Uninsulated hot surface heat loss _____ (@ 3 btu/ft²-ΔT)
10. Electrical disconnect/e-stop installed _____Yes _____No

Feed Water Tank Design Considerations

1. Feed water tank working volume _____gallons
2. Feed water tank full volume _____gallons
3. Tank stand height _____ft above level floor
4. Tank construction _____ (SS, CS)_____gauge _____ × _____ size
5. DA type_____ (spray or tray)_____ .005 cc/l or .05 cc/l _____Atmospheric
6. Tank operating level controls type _____
7. Tank low water cutoff control type _____
8. Tank preheat system type _____ sized for _____lb/h steam (kW)
9. Makeup water source _____ maximum M/U rate _____gpm
10. Number of feed water pumps_____ type _____
11. Pumps _____on/off _____ continuous run _____ VFD motors
12. NPSHA calculated _____
13. Pump flow rate _____gpm @ _____psig with _____ft NPSHR
14. Chemical quills installed _____ (1 or 2) location _____on the tank
15. _____sight glass _____magnesium anode _____ tank insulation

Blowdown System Considerations

1. _____Blowoff separator included _____ Model _____
2. ASME rating _____ Construction CS/SS _____
3. Cooling kit _____automatic TCV _____ manual valve
4. _____sight glass installed on BO separator
5. Boiler bottom blowdown valves _____type _____number

6. _____Surface blowdown system installed
7. Surface BD _____ automatic or _____ manual
8. _____conductivity based _____ timer based
9. Boiler EER calculation _____
10. Boiler COC calculation_____
11. Boiler total BD rate to meet COC _____

Steam Delivery System Considerations

1. Steam main size _____ yields _____ft/min steam flow
2. Multiple boiler high water protection method _____
3. _____PRV station _____ pressure to _____ pressure
4. _____model PRV _____size PRV
5. _____ No. air vents installed _____
6. _____ Drip legs installed one every _____feet
7. Steam piping insulation spec _____
8. Type of control valves _____

Condensate Collection System Considerations

1. Condensate line sizes_____
2. Steam trap type/sizes _____
3. _____pump traps _____ model, Air/steam pressure _____psig
4. _____electric condensate collector _____pumps
5. Surge tank size _____×_____ construction CS/SS _____ gauge _____ASME
6. ST pumps _____ Model_____ Capacity _____gpm @ _____TDH
7. ST level control system_____
8. ST fresh makeup rate estimate _____gpm

Water Treatment Considerations

1. Corrosion protection scheme _____
2. Fouling protection scheme _____
3. Scaling protection scheme _____
4. No. chemical feed systems_____ Tank sizes _____
5. Pump rates_____gph @_____TDH, Model_____

6. Makeup water rate _____gpm _____ F temperature
7. Makeup water treatment scheme _____
8. RO size _____ gpm with _____ gpm permeate _____gpm concentrate
9. Softener sizing _____ Duplex or simplex_____ M/U water _____gpg
10. Softener grain per day removal calculation _____
11. Softener tank sizes _____ Brine storage tank size _____
12. Softener inlet water connection size _____ inch

D

PROBLEM SET ANSWERS

Answer 1. A DA tank that is $36'' \times 60''$ will have a surface area around $2\pi R \times L + (2 \times \pi D)$ or $2 \times 3.14 \times 1.5$ ft $\times 5$ ft $+ (2 \times 3.14 \times 3$ ft$) = 65.94$ or 66 ft². At 5 psig the tank contents and surface will be 227 F. Therefore the temperature difference (ΔT) between the DA and ambient air is 227 F – 75 F or 152 F. Table 3.4 indicates a horizontal cylinder with natural convection will radiate about 313 btu/ft²-h. At 66 ft² this tank will lose 66 ft² $\times$ 313 btu/ft² -h = **20,658 btu/h**.

Conversely, you could use the general heat transfer equation to solve. $Q = m \times U \times \Delta T$, where $m = 66$ ft², $U = 2.5$ btu/ft²-h-F (from Table 3.3), and $\Delta T = 152$ F. This method indicated about **25,000 btu/h are lost**.

If the tank is insulated with 2 inches of glass wool insulation, the rule of thumb is the heat loss will be reduced by at least 90%. Therefore, insulating this tank would reduce the heat loss by $20,658 \times 10\% = $ **2066 btu/h**. Powerful case for energy savings.

If the ambient airflow is increased to 10 FPS near this tank, then the uninsulated tank heat loss would be around 66 ft² $\times$ 714 btu/ft²-h (Table 3.4) = **47,124 btu/h**.

Answer 2. The extraction steam would need to increase the makeup water rate from 50 to 205 F. We also know that a 100 hp steam boiler will produce steam at a rate of 0.069 gpm/BHP $\times$ 8.34 lb/gal $\times$ 100 BHP = 57.5 lb/min. The energy required to raise 57.5 lb of water from 50 to 205 F is 57.5 lb/min $\times$ 1 btu/lb $\times$ (205 F – 50 F) = 8912 btu/min or 534,750 btu/h. 120 psig steam will provide about 872 btu/lb of latent heat; therefore, the extraction steam required will be about **613 lb/h**.

Process Steam Systems: A Practical Guide for Operators, Maintainers, Designers, and Educators, Second Edition. Carey Merritt.
© 2023 John Wiley & Sons, Inc. Published 2023 by John Wiley & Sons, Inc.

Answer 3. The amount of heat loss from this pipe will depend on the ambient room temperature and the air flow velocity around the pipe. The air flow is stated as convection only, and if we assume 70 F room temperature, then the values in Table 13.2 can be used to estimate heat loss. Table 13.2 shows an eight inch pipe carrying 10 psig steam (239 F) will lose about 906 btu/h-ft. Therefore, the 750 ft of steam line will lose 750 ft × 906 btu/hour-ft = **679,500 btu/h**.

Alternatively, you could determine the total area of the 750 ft pipe and use the general heat transfer equation to estimate heat loss. $Q = mCp\Delta T$. Where $m = 750$ ft × 2.258 ft²/ft (for 8″ pipe), Cp = 2.5 btu/ft, and $\Delta T = 239$ F – 70 F. Solving for Q, the heat loss is **715,503 btu/h**.

Answer 4. A 400hp steam boiler will have an input of about 13.8 MMbtu/h. The radiant losses will then be 13.8 MM × 0.02 = 276,000 btu/h. These losses are the same regardless of the firing rate. **At high fire (100% output) this represents a 2% loss in efficiency. However, at 20% firing rate the efficiency becomes 276,000 btu/2.76 MMbtu = 10%**. This is why a boiler operating at low fire for extended periods of time should be replaced with a smaller boiler that will operate at a much higher firing rate.

Answer 5. At 100 psig the blowdown boiler water will flash off at a rate of 13.3% in an atmospheric blowoff separator. Therefore, the amount of flash steam will be 12 gal × 8.34 lb/gal × 0.133 = **13.3lb/blowdown**.

If the blowdown occurs in 30 s, then this equates to 12 gal/30 s × 60 s/min = 24 gal/min. With 13.3% flash steam, the equivalent mass flow out the separator vent would be 26.6 lb/min. A 0 psig, steam has a specific volume of 26.8 ft³/lb. Therefore, the 26.6 lb/min of flash steam will have a discharge flow rate of 26.6 lb/min × 26.8 ft³/lb = 713 ft³/min. A 3″schedule 40 vent line has a volume of 0.05135 ft³/ft. Therefore, the 3″ vent pipe will have a steam velocity of 713 ft³/min/0.05135 ft³/ft = **13,885 ft/min**.

The amount of recoverable energy from the blowdown water will depend on what is being heated with waste heat from the boiler water. We know that saturated water (from saturated water tables) at 100 psig will have an enthalpy of 310 btu/lb. If we heat domestic water up to 120 F, then the recoverable heat would be determined by the enthalpy difference between the saturated boiler water and the 120 F water. The saturated water charts show 120 F water to have an enthalpy of 88 btu/lb. Consequently, the blowdown boiler water could yield a maximum energy recovery of 12 gal/blowdown × 8.34 lb/gal × 1 btu/lb × 310 – 120 F = **19,015 btu/blowdown**. The actual recoverable heat could only be calculated based on the heated water flow rate and heat exchanger performance used to transfer the energy.

Answer 6. The combustion efficiency charts in Appendix A show a gas-fired boiler with a net stack temperature of 380 F – 80 F = 300 F and a CO_2 level of 8% will have a combustion efficiency of 100 – 16.7% = **83.3%**.

The same boiler operating a high fire with a net stack temperature of 490 F – 80 F = 410 and a CO_2 level of 9.8% will have a combustion efficiency of 100 – 17.6 = **82.4%**.

Answer 7. Combustion efficiency is a measure of the performance of the burner and pressure vessel as it relates to the fuel/air combustion and the heat transfer of that energy to the boiler water. Fuel to steam efficiency subtracts the radiant heat losses and blowdown energy losses from the boiler combustion efficiency. Fuel to steam efficiency measures the difference in fuel energy input to the steam energy output.

Answer 8. The answer to this question is related to latent heat. Since most steam systems that use heat exchangers are designed to transfer only latent heat, then low pressure steam will transfer more heat per pound than high-pressure steam. For example, **12 psig steam has latent heat of 950.1 btu/lb, and 120 psig steam has latent heat of 871.5 btu/lb**.

Answer 9. You will need a steam supply with at least 10–20 F higher temperature than the temperature the material you wish to heat. Therefore, if you wish to heat a material up to 350 F, you would need a saturated steam supply having a temperature of 140–160 psi steam. You could also use a steam with a lower pressure, but it would need to be superheated to at least 360–370 F.

The material you wish to heat up has a heat capacity of 1.2 btu/lb-F, weighs 1000 lb, and is at 70 F. You will need 1000 lb × 1.2 btu/lb × (350 – 70)F = 336,000 btu to heat this material from 70 to 350 F. If the heat up rate is 30 min, then you would need a steam supply capable of transferring heat energy at a rate of 336,000 btu/0.5 h = **672,000 btu/h**.

Answer 10. To determine the amount of steam required to heat this solution you will need to know the heat capacity of 10% sulfuric acid (SA), and the latent heat of 50 psig steam. You find a 10% sulfuric acid solution has a heat capacity of 0.9345 btu/lb-F and 50 psig has latent heat equal to 912.2 btu/lb. Therefore, to heat 20,000 lb of 10% SA × 0.9345 btu/lb-f × (180 – 80)F = 1,869,000 btu. For a 2 h heat up period we then will need 1,869,000 btu/2 h = 934,500 btu/h steam heat. If we use 50 psig steam we can see that the amount of steam required is 934,000 btu/h/912.2 btu/lb = **1024 lb/h of 50 psig** steam. This calculation does not account for any surface heat losses. Therefore the tank size or openness information is immaterial.

The heat required to maintain the tank at 180 F can be calculated by determining the heat required to heat anything added to the tank and determining the heat loss by convection from the open top. If we dip 4000 lb of 316L stainless steel into this tank three times an hour, we will need to heat 12,000 lb of this steel from 80 to 180 F. 316L stainless steel has a heat capacity of about 0.4 btu/lb-F. Therefore the heat required to heat the steel per hour is 12,000 lb × 0.4 btu/lb-F × (180 – 80)F = **480,000 btu/h**.

In addition, the open top tank will lose some heat via convection. From Table 3.4 there is data that will help determine this heat loss; however, you must interpolate twice. First find convection only and 10 FPS air flow from a horizontal surface at 180 F. By interpolating 150 and 200 F values you will get 150 + (240 – 125) × 60% = 219 btu/SF-h at 180 F with convection only. Likewise, at 10 FPS and 180 F the heat loss is 312 + (565 – 312) × 60% = 464 btu/SF-h. Almost there. Now the value of heat loss at 5 FPS air flow will be somewhere between convection losses and 10 FPS air flow losses. For this we can take the mean and be relatively close to actual heat loss. Therefore, an 8′ × 24′ open top tank has 192 SF surface area and will lose (219 + 464)/2 = 341 btu/SF -h × 192 SF = **65,472 btu/h**. The answer to the question is the 65,472 btu/h + 480,000 btu/h = **536,472 btu/h**.

Answer 11. Any time a boiler that is rated for a high pressure is operated at a lower pressure, you must consider the consequences of steam nozzle velocity. A 150 hp, 150 psig steam boiler with a 3 inch Schedule 40 steam nozzle will have a steam velocity of <5000 FPS, which is considered acceptable. However, at 20 psig the steam nozzle would need to be 8 inches to keep exit velocity less than 5000 FPS. **The small steam nozzle will likely create excessive steam velocities and poor steam quality**.

Should you encounter such a problem, you could add a steam moisture separator on the steam header to remove excess carryover boiler water. The removed water should be routed back to the boiler or to the feed water tank.

Answer 12. The impact on this plant's steam system can occur by the instantaneous heating of the water and by the daily water usage rate. Therefore, both usages must be determined. Since we are interested in the maximum steam use, we should use the lowest water supply temperature of 45 F. Then heating 20 gpm of 45 F water up to 140 F will require 20 gal/min × 8.34 lb/gal × 1 btu/lb × (140 – 45)F = 15,846 btu/min. This equates to an hourly steam flow of 15,846 btu/min × 60 min/h/905 btu/lb of steam = **1050 lb/h steam use rate**. Furthermore, at 1050 lb/h steam usage rate, this equates to an equivalent BHP of 1050 lb/h/34.5 lb/h/BHP = **30.4 boiler horsepower**.

You will also need to know the daily steam use. Therefore, the 6000 gallon/day steam heat use will be 6000 gal/day × 8.34 lb/gal × 1 btu/lb × (140 – 45)F = 4,753,800 btu/day or 4,753,800 btu/day/905 btu/lb = **5253 lb steam/day**. The actual usage per hour would depend on when the water is heated. This is why some industries use house steam to heat up water at night (no manufacturing going on) and store the hot water for day-time use.

Answer 13. If the boiler feed water is 325 us/cm, and the boiler water is 2500 us/cm, the calculated cycles of concentration in the boiler is 2500/325 = 7.7 cycles of concentration (COC). The blowdown rate (BD) will be equal to EER/1 – COC where EER = bhp × 0.069gpm/bhp × (100% – condensate return percent). Therefore a 300

BHP boiler with 50% condensate will have an EER equal to 300 bhp × 0.069 gpm/bhp × 50% (or 0.5) = 10.35 gpm. **Blowdown rate is then 10.35 gpm/ 1 – 7.7 = 1.54 gpm.**

Answer 14. ACFM is the actual flue gas flow at the stack temperature. A 600 bhp burner at rated input of 25 MMbtu natural gas will have an input of 25,000,000 btu/h/1000 btu/CF natural gas/60 min/h = at 20% excess air will use 417 CFM of natural gas. Since natural gas burns at a 1/10 ratio of natural gas to air, the burner input is then 417 CFM gas plus 4170 CFM air. Furthermore, with 20% excess air the total air to the burner is then 4170 × 1.20 = 5004 CFM air. The burner input is then 417 CFM gas plus 5004 CFM air or 5421 CFM fuel mixture. This calculation is based on ambient temperatures which can be assumed to be near 25C (77 F). This flow rate of **5421 CFM is the standard flow rate or SCFM** of the fuel mixture before being combusted.

The actual stack flow rate (ACFM) will account for the temperature difference between the SCFM value and the actual stack temperature of 420 F. To calculate ACFM use the ideal gas law $PV = NRT$. Since pressure (P), n, and R are equal for these conditions, the formula reduces to $V_1/T_1 = V_2/T_2$. Solving for V_2 we get $V_2 = V_1 \times T_2/T_1$. So, V_2 = 5421 cfm × 215.5 C (420 F)/25 = **46,729 ACFM**.

To model the stack static pressure you will need to know stack diameter, length of horizontal run, and length of vertical run and ambient outside design temperature.

A good stack static pressure is –.02 inches of water column under all ambient conditions.

Answer 15. When steam passes through a pressure reducing valve its enthalpy remains constant but pressure is lower. Looking at the saturated steam table we see that saturated 130 psig steam has an enthalpy of 1194 btu/lb and at 60 psig has a total enthalpy of 1182 btu/lb. From the data we can see the lower pressure steam will then have an additional 12 btu/lb energy which makes it **superheated steam**.

Interestingly, the higher pressure steam has a latent heat of 868 btu.lb and sensible heat of 327.5 btu/lb. When the steam passes through the PRV, its latent heat rises to **905.3 btu/lb and its sensible heat will be 1194 – 905.3 btu/lb = 288.7 btu/lb**. Even though the lower pressure steam is superheated, it has a lower sensible heat value.

Answer 16. A feed water pump must supply water to the boiler at a rate higher that the evaporation rate and a higher pressure to adequately keep the boiler filled. Boiler code states the minimum feed pump discharge pressure must be at least 105% of the boilers trim pressure. Furthermore, for on/off feed water pump controls, the rule of thumb is a flow rate equal to about 2.5 times the boilers rated steam output.

A 100 hp boiler trimmed at 150 psig should have an on/off feed water pump capable of pumping 100 bhp × 0.069 gpm/bhp × 2.5 = **17.25 gpm and 105% × 150 psig or 157.5 psig**.

Answer 17. We know that oxygen solubility in water is inversely proportional to the temperature of the water. Using the chart found in Figure 10.2 we can see the oxygen concentration at 120 F will be about 5.5 ppm, and at 200 F the oxygen concentration will be about 1.4 ppm. The remaining oxygen can be removed by either heating the water up closer to the boiling point or by adding an oxygen scavenging chemical like sodium sulfite.

Answer 18. Table 13.3 shows steam flow through a ¼″ orifice at 50 psig will pass 24.5 lb/h. This would equate to a loss of 24.5 lb/h × 24 h/day × 30 days/month = **17,640 lb/month**.

Answer 19. When steam condenses it entraps carbon dioxide which makes the condensate slightly acidic. The low pH condensate will aggressively corrode mild carbon steel. This is why condensate piping is generally made from **stainless steel**. Acidic condensate can be prevented by adding a **neutralizing amine** to the boiler water or steam header. The amine will react with the carbon dioxide and essentially remove it from the condensate.

Answer 20. Water contains hardness ions such as calcium and magnesium. A water softener contains resin beads with hundreds of thousands of little charged sites that attract calcium and magnesium ions. Virgin resin sites have a sodium or hydrogen ion that has a lesser affinity than the hardness ions. Consequently, when the hard water is passed through the softener resin, the hardness ions stick to the resin and the sodium or hydrogen ions fall off and remain in the water.

Once the resin sites become full of hardness ions (considered spent resin), the resin sites can be regenerated using a very high concentration of salt (sodium chloride). This is what we use the salt for. As the salt solution passes through the spent resin the hardness ions are replaced with new sodium ions. This occurs for softener resin that uses sodium based resin. If the softener used hydrogen based resin, the regenerate solution would be a strong acid solution (source of H+ ions) instead of salt.

Answer 21. The purpose of a flash tank is convert pressurized condensate into atmospheric condensate. Pressurized condensate has a much higher potential to create steam hammer and condensate pumping problems. Many condensate system do not require flash tanks (i.e., low pressure steam systems). A blowoff separator is essentially a flash tank as it takes hot pressurized boiler water and converts it to atmospheric hot water.

Answer 22. When a boiler is turned to the on position a series of checks are automatically performed to make sure the boiler system is safe and ready for fuel combustion. The following are system parameters that would prevent fuel flow to the burner or cause a failure.

a) Fuel pressure to high or too low.
b) Air flow is not high enough through the burner.
c) Limit switches on the inlet air damper are not set right to release the programmer to initiate combustion.
d) In addition, the fuel rod or other ignition source in the burner could be faulty preventing ignition.

Answer 23. You will find this information in the saturated steam and saturated water charts. The amount of energy the steam gives up can be calculated by subtracting the enthalpy of 180 water from 125 psig steam. The charts show 180 water contains 148 btu/lb and 125 psig steam contains 1193 btu/lb. Therefore, 600 lb of steam cooled to 180 F water will release 600 lb × (1193 – 148) btu/lb = **607,800 btu.**

Answer 24. We know natural gas will require at least a 10:1 air to gas volume mixture for ideal combustion; however, excess air is added to facilitate complete combustion. Therefore, 2000 CF of natural gas using 20% excess air will require (20,000 CF air) × 1.2 = **24,000 CF air**.

Answer 25. Two 750 hp steam boilers will use 0.069 gpm/bhp × 750 bhp × 2 = 103.5 gpm of feed water at 100% output. At 75% output they require 103.5 gpm × 0.75 = 77.6 gpm. Furthermore, with 75% returns, the makeup rate will be 25% of 77.6 gpm or 77.6 gpm × 0.25 = **19.4 gpm**.

The combined condensate and makeup water temperature will be ((77.6 gpm × 0.75) × 195 F + (77.6 gpm × 0.25) × 65 F) 77.6 gpm = **162.3 F.**

The amount of steam required to heat the feed water water up to 227 F can be determined by subtracting the enthalpy of 77.6 gpm at 162.3 F from the enthalpy of 77.6 gpm at 227 F water. From the saturated water charts we can that 162.3 F water has an enthalpy of 130.2 btu/lb, and water at 227 F has an enthalpy of about 195.2 btu/lb. The amount of energy required to heat the 77.6 gpm up to 227 F is then 77.6 gpm × 8.34 lb/gal × (195.2 – 130.2) btu/lb = 42,067 btu/min. Furthermore, steam at 120 psig has an enthalpy of 1193.3 btu/lb. Therefore, the amount of steam required to add 42,067 btu/min can be divided by 1193.3 btu/lb which **equals 35.3 lb of steam per minute or 2115 lb/h**.

Note: The injected steam will displace some makeup water (35.3 lb/min/8.34 lb/gal = 4.2 gal/min) to the DA and the steam required to heat up the 77.6 gpm feed water will be slightly less.

A properly operating DA that has 227 F water will be around 5 psig pressure.

Answer 26. Since 200 psig steam will only be 387.7 F, you will need to use super-heated steam and a heat exchanger to heat this material. Since the desired material temperature is 620 F, the steam heating source should be at least 10% higher temperature. We can use 700 F as the steam temperature required. 2400 lb of this material will require 2400 lb × 0.8 btu/lb-F × (620 − 100)F = **998,400 btu added to the material**.

200 psig saturated steam has a total heat of 1200 btu/lb. The superheated steam charts show 200 psig steam at 700 F will have an enthalpy of 1373 btu/lb. An additional **173 btu lb** of energy is required to superheat the 200 psig steam to 700 F.

The amount of energy required to superheat the steam can be determined by subtracting the energy of the material at 387.7 F from the energy at 620 F. Therefore, 2400 lb × 0.8 btu/lb-F × (620 − 387.7)F = **446,016 btu** is the amount of superheat energy required to be added to the steam.

We can determine the amount of 200 psig steam required for this application by determining the amount of steam needed to raise the temperature of the 2400 lb from 100 to 387.7 F. This works out to be 2400 lb × 0.8 btu/lb-F × (387.7 − 100)F/838.4 btu/lb = 659 lb of steam. To raise the 446,016 btu is the amount of additional super-heat energy that is needed.

The conversion of thermal energy to electrical energy is 3412 btu/kW. Therefore the amount of electrical energy required to superheat is then 446,016 btu/3412 btu/kW = **130.7 kW**.

Answer 27. To determine the water treatment capacity of the water softener we must first convert the 250 ppm concentration value to grains/gallon of hardness. From the chapter on chemistry control we know that 17.1 ppm equals 1 grain/gallon (gpg). Therefore a water softener that has a capacity of 150,000 grains of hardness will be able to soften 150,000 grains/(250 ppm/17.1 grains/ppm) = **10,260 gallons** of 250 ppm hardness water.

Answer 28. Linear expansion of a metal pipe will be the same regardless of pipe size; however, will be different for different metallurgies. As a rule, carbon steel will expand to a lesser degree than stainless steel for the same temperature differential. The formula for linear expansion shown in Chapter 9 is, Expansion (in) = Length (in) × Delta Temperature(F) × (α) Coefficient of linear expansion for that metal. The α values for A36 carbon steel are 0.0000065 in/in-F and for 304SS is 0.0000096in/in-F. Delta temperature is 70 F − 350.1 F or 280.1 F. The linear expansion of 1200 ft of steam line for each metal is then

For A36: 1200 ft × 12 in/ft 0.0000065 in/ft-F × 280.1 F = 26.2 inches

For 304SS: 1200 ft × 0.000096 in/ft-F × 280.1 F = 38.7 inches.

The expansion needs to accounted for in the design. Expansion loops and pipe mounting that allow for lateral movement is essential.

Answer 29. Water hammer is caused by a slug of water that moving rapidly through piping and then hits an obstruction (i.e., closed valve, tee, or 90 elbow). The momentum stop of the water causes a high-pressure wave at the point of impact. The pipe can jump violently or fail. Water hammer in steam headers is caused by the buildup of condensate that forms a fast moving slug of water. Water hammer in steam lines can be mitigated by proper draining via drip legs and maintaining functioning steam traps. Water hammer in water lines can be mitigated by slowing down the opening/closing of valves, use of slow starts on pumps, or by the use of surge suppressors.

Answer 30. Steam hammer is a result of thermal shock. When steam contacts a cool pipe or cool media, the steam can violently collapse creating a massive negative pressure spike. This sudden pressure spike can result a "banging" or hammering noise. Steam flashed into cool condensate lines can result in steam hammer and can be mitigated by use of flash tanks or condensate diffusers.

Answer 31. A boiler's ability to respond to sudden changes in load demand is related to its water volume. Generally the higher the water volume the better the boiler will handle such demand. The consequences of load sudden demand is a rapid depressurization of the boiler. When this rapid depressurization occurs, some of the boiler water will flash off steam. Consequently, the higher water volume produces more flash steam and prolongs the ability to provide steam to the steam header.

Answer 32. The NOx concentration is usually standardized with emissions data corrected for 3% O_2. The formula for calculating NOx corrected is equal to NOx measured × (10.23%/measured CO_2%). Therefore, the NOx at 3% O_2 is 34 ppm × (10.23% − 8.4%) = **62.2 ppm**.

Answer 33. Not all steam systems will require the same level of gauges and instrumentation. Therefore, I recommend the following sets. For simple small scale systems that supply only one piece of equipment or one zone you should include the basic set of gauges and sensors. Boiler pressure gauge, feed water tank water level gauges, feed tank and boiler stack temperature gauges, feed water pump discharge gauge, and fuel train pressure gauge. All boiler must have pressure sensors and a water level gauge.

For more complex steam systems I recommend economizer inlet and outlet temperature gauges, steam header pressure sensors, feedwater temperature gauges, stack draft pressure sensor, makeup water and steam flow meters, fuel inlet and last elbow pressure sensors, and boiler blowdown conductivity meter.

Answer 34. Most liquid and gas regulators, sensors, and meters require laminar flow to yield consistent output readings and control function. The upstream and downstream clear pipe run requirements provide minimum distances that will yield laminar flow. These are minimum distances and failure to abide by these recommendations will usually result in poor performance and premature equipment failures.

Answer 35. Any steam (and other gas or liquid) sensing equipment will provide the most accurate and steady measurement when subjected to laminar flow of material. The 10 pipe diameters refer to a minimum distance any turbulent flow will become laminar. The 10 pipe diameters is a minimum distance; more is always encouraged.

INDEX

Process Steam Systems: A Practical Guide for Operators, Maintainers, Designers, and Educators,
Second Edition. Carey Merritt.
© 2023 John Wiley & Sons, Inc. Published 2023 by John Wiley & Sons, Inc.

Printed and bound by CPI Group (UK) Ltd, Croydon, CR0 4YY

16/04/2025

14658415-0002